A Splintered Vision

D1160387

A Splintered Vision

An Investigation of U.S. Science and Mathematics Education

WILLIAM H. SCHMIDT
Michigan State University, East Lansing, Michigan, U.S.A.

CURTIS C. MCKNIGHT
University of Oklahoma

and

SENTA A. RAIZEN
The National Center for Improving Science Education

With the collaboration of

Pamela M. Jakwerth
Gilbert A. Valverde
Richard G. Wolfe
Edward D. Britton
Leonard J. Bianchi
Richard T. Houang

of the
U.S. National Research Center for the
Third International Mathematics and Science Study

KLUWER ACADEMIC PUBLISHERS
DORDRECHT / BOSTON / LONDON

A C.I.P. Catalogue record for this book is available from the Library of Congress.

ISBN 0-7923-4440-5 (Volume 3)
ISBN 0-7923-4442-1 (Set of 3 Volumes)
ISBN 0-7923-4441-3 (Volume 3 PB)
ISBN 0-7923-4443-X (Set of 3 Volumes PB)

Published by Kluwer Academic Publishers,
P.O. Box 17, 3300 AA Dordrecht, The Netherlands.

Kluwer Academic Publishers incorporates
the publishing programmes of
D. Reidel, Martinus Nijhoff, Dr W. Junk and MTP Press.

Sold and distributed in the U.S.A. and Canada
by Kluwer Academic Publishers,
101 Philip Drive, Norwell, MA 02061, U.S.A.

In all other countries, sold and distributed
by Kluwer Academic Publishers,
P.O. Box 322, 3300 AH Dordrecht, The Netherlands.

03-0298-250 ts

Printed on acid-free paper

All Rights Reserved
© 1997 Kluwer Academic Publishers
No part of the material protected by this copyright notice may be reproduced or
utilized in any form or by any means, electronic or mechanical,
including photocopying, recording or by any information storage and
retrieval system, without written permission from the copyright owner.

Printed in the Netherlands

Dedication

To Carrie and Jason whose school experiences made me aware of the importance of educational opportunity and made me care about its implications in children's lives.

W. H. Schmidt

To Robert B. Davis and Kenneth J. Travers, my mentors in mathematics education and cross-national studies respectively — who taught me how important it is to understand what actually happens in classrooms before recommending changes.

C. C. McKnight

Acknowledgments

Our vision for this manuscript would have remained splintered without the support and insight of numerous friends and colleagues. We are indebted to the countless visions we encountered throughout the production of this manuscript and would like to express our gratitude and appreciation for the generosity displayed by many. While the substance of *A Splintered Vision* reflects the perspectives of the authors alone, this publication would not have been possible without the participation of the following:

- Albert Beaton, the Third International Mathematics and Science Study director, has provided general oversight and direction to the many different components of the TIMSS research project.

- The U.S. Steering Committee for TIMSS - Gordon Ambach, Deborah Ball, Audrey Champagne, Jewell Plummer Cobb, David Cohen, John Dossey, Emerson Elliott, Sheldon Glashow, Larry Hedges, Henry Heikkinen, Jeremy Kilpatrick, Mary Lindquist, Marcia Linn, Robert Linn, Paul Sally, Richard Shavelson, Bruce Spencer, Elizabeth Stage, James Taylor, Kenneth Travers, and Paul Williams - helped guide U.S. participation in TIMSS. Many members also reviewed early drafts of this manuscript and provided invaluable perspectives on the ideas expressed here.

- Members of the U.S. National Coordinating Committee - Jeanne Griffith, Eugene Owen, Lois Peak, Larry Suter - helped to coordinate various aspects of the project and offered guiding comments on our work at all stages.

- Rodger Bybee, Michael Kirst, Barbara Plake, Richard Prawat, Thomas Romberg, and Laurence Wolff were sent review copies of this manuscript, and the feedback received furthered our work.

- Haiming Hou, Wen-Ling Yang, Christine deMars, and Shelly Naud are responsible for much of the statistical work behind many of the exhibits. Our work would not have been possible without the results of their analyses.

- Jacqueline Babcock, Leland Cogan, Emilie Curtis, Mike Reed, and Lorene Tomlin contributed countless hours in completing the many mundane and tedious tasks behind the production of this publication. We would be lost without them.

Finally, the work reported here would not have been possible without the support of grants RED 9252935, SED 9054619, and RED 9550107 from the National Science Foundation in conjunction with the National Center for Educational Statistics. NSF and NCES provided the funds for this resarch, but do not assume responsibility for the findings or their interpretations.

We are grateful for all who helped shape and focus our vision.

Table of Contents

INTRODUCTION

There is no one at the helm of U.S. mathematics and science education. In truth, there is no one helm. No single coherent vision of how to educate today's children dominates U.S. educational practice in either science or mathematics. There is no single, commonly accepted place to turn to for such visions. The visions that shape U.S. mathematics and science education are splintered. This is seen in what is planned to be taught, what is in textbooks, and what teachers teach.

The above paragraph states some of the overall conclusions drawn in the present report. Their more detailed forms are discussed further below and data supporting them are presented. This report discusses data from the analysis of 628 textbooks and 491 curriculum guides from around the world as a part of the recently completed curriculum analyses of the Third International Mathematics and Science Study (TIMSS).* It also presents accompanying data on teacher practices in the U.S. and two other countries (since the full TIMSS teacher data set is not available for release at this point).

---

*TIMSS is a large-scale, cross-national comparative study of several aspects of national educational systems and their outputs. It involves gathering data on curricula in mathematics and the sciences, data on instructional practices, and data on school and social factors, as well as student achievement testing at selected populations. About 50 countries are involved in one or more aspects of TIMSS. The TIMSS curriculum analysis gathered data using a variety of methods. These included gathering representative documents that laid out official curricular intentions and plans, collecting and analyzing entire mathematics and science textbooks in representative samples at the grades in which achievement testing would also take place, and analyzing entire K-12 textbook series for selected "in-depth" topics. These data were supplemented by data provided on the placement in the curriculum of subject-matter sub-areas across the grades and by questionnaire data on the structure of educational systems in each country and national experts' opinions on a few central issues such as planned or on-going reforms. The document analyses of textbooks and curriculum guides used common "frameworks" of subject-matter sub-areas and expected student capacities. Each document was partitioned first into larger "units" and then further divided into small, single-purpose sub-units we called "blocks." Each block was characterized by its type, subject-matter content, expected student performances, and other aspects. This partitioning and characterization was done by coders in each participating country who had been trained by trainers themselves taught in regional meetings. Coders' work was subject to quality assurance procedures throughout the coding and the resulting quality and reliability assessed after the data were received. All data were audited and ambiguities clarified with the participating countries. The result is a massive base of largely document-based curricular data from almost 50 countries that has been gathered, created, and cleaned under strict controls and has had its quality assessed both centrally and by each country's audits of their own data after entry into that central data base. The current report draws heavily on the curriculum documents and textbooks analyzed as part of the U.S.'s participation in that aspect of TIMSS.

Other TIMSS reports supply more details on these matters and others. The TIMSS curriculum and teacher data are extensive and cannot be explored in a single report. The results of analyses of these data are being reported in a series of volumes; three of which are now available. The first, *Characterizing Pedagogical Flow,*[1] discusses curriculum data in mathematics and science along with classroom observations and teacher interviews in six TIMSS countries. The second and third, *Many Visions, Many Aims: A Cross-National Investigation of Curricular Intentions in School Mathematics* and *Many Visions, Many Aims: A Cross-National Investigation of Curricular Intentions in Science Education,*[2] are reports that present data on the full set of almost 50 TIMSS countries for several aspects of curricula in mathematics and science education.

The present volume is the first to investigate these results in more detail for the United States, to compare them to data on teacher practices, and to interpret the findings and encourage their discussion among those concerned about science and mathematics education in this country. Other technical and interpretive volumes will follow.

The present report intends to document and characterize the state of U.S. mathematics and science curricula and place them in a cross-national context. Fortunately, educational practice in the U.S. is a "moving target." These data were collected in 1992-93. At that time the National Council of Teachers of Mathematics (NCTM) *Standards*[3] (for mathematics education) had only existed for about three years. The American Association for the Advancement of Science's (AAAS) *Benchmarks*[4] (for science and mathematics literacy) had been released only in preliminary form. The National Academy of Science's National Research Council's[*] *Science Education Standards*[5] had yet to be fully formulated or released. The intervening years have been a time of change for state curriculum standards[6] and textbooks.

Have mathematics and science education's curricular visions and textbooks changed so quickly that the current data are no longer valid? The TIMSS data on teacher practices discussed here were collected in 1995. The fact that these data agree so well with earlier data on curricula and textbooks suggests that, whatever changes occur at the level of "intended" (planned for, envisioned) curricula, changes in "implemented" (actual classroom practice) curricula lag somewhat behind (assuming, perhaps optimistically, that what we intend in time affects what is practiced).

The data here characterize central aspects of this moving target of U.S. science and mathematics educational practice. They are a snapshot rather than a moving picture. To the extent that change permeates educational systems in our country over a period of years, these data reveal something of what is. At worst, they raise questions by picturing what recently was true and

[*]The *Science Standards,* although issued by the National Academy of Science, were the product of working groups including teachers, science educators and other scientists. The same was true for the NCTM *Standards* and, to some extent, for the AAAS *Benchmarks.* While the development of these documents was fostered by the issuing organization, they represent outcomes from a range of diverse inputs.

what yet may be true, and they highlight the need for a more current look at mathematics and science curricula and textbooks to see just how much change has occurred in the past three years.

This report is meant to be descriptive and, to a lesser extent, interpretive. It is not a direct call for specific reforms. We seek to provide data germane to the ongoing public debate about science and mathematics education policies in the U.S. The data seem at times to demand changes and improvement. If so, the data themselves, rather than our comments, should spur important discussion about parts of our national educational future. If this discussion of our data occurs, this report will have done its job.

The Splintered Vision: An Overview

Guiding visions in science and mathematics education splinter partly by necessity. Our national visions are composites of visions articulated in thousands of essentially independent official sites — local, state, and federal. Official voices shape their own guiding visions in a running "conversation" with textbook publishers, test writers, professional organizations, and others who contribute to the ongoing dialogue in education.

Education in the U.S. has always been guided by agencies and organizations — local, state, and national — that each take their share of responsibility for education. Our earlier statement that there was "no one helm" for mathematics and science education should not be taken as implying that there should be. There can be strength in such diversity. At its best, our system of distributed educational responsibility allows local preferences and community needs to help determine what occurs in local schools. It also provides laboratories in which varied approaches are implemented and then used to inform the others, in an effort to determine more effective educational practices. Unfortunately, there also can be weakness in this decentralized responsibility. At its worst, our system requires that we seek consensus on needed changes site by site. It is an open question whether this fragmenting due to the distribution of responsibility helps create other, less desirable kinds of fragmentation in mathematics and science curricula. The serious consequences of our sharing responsibilities for education are part of what the data presented here may partially illuminate.

In any case, our educational "system" implies that any desired changes in mathematics and science education must be implemented within our context of shared educational responsibility. A corollary is that we may learn from other countries but we cannot emulate their centrally administered changes. Any reform in the U.S. must seek visions that can achieve broad consensus.

Unfortunately, the nature of our system is not the only source for splintered educational visions. Other forces are at work. Each site's vision fragments further into smaller visions — pieces that often do not coalesce into a well-guided whole. Our data strongly suggest that these many splintered versions and pieces of visions do not form an intellectually coherent vision to guide our policies and actions in providing the best for our children learning mathematics and science. An overview of our empirical findings are presented in the rest of this section.

Official curricula. Our data show that splintering manifests itself in curricular intentions in many ways. For example, we found that

- States in the U.S., on average, indicated plans to cover so many topics* that the composite of intended topics included more topics until the ninth (mathematics) or tenth (science) grade than 50 - 75 percent of the other countries studied.
- The composite U.S. curricula in mathematics and science show that we planned to add as many new topics as was typical in other countries, but we did not drop topics until well into lower and upper secondary school.
- Science and mathematics topics remained in our composite U.S. curricula for more grades than all but a few other TIMSS countries.
- While the core of mathematics topics was broad, it varied little among the states. On the other hand, the number of core science topics was much smaller, and the overlap among state curricula was also small. While students in U.S. states might have studied a number of science topics roughly equal to the international median, the students in different states likely studied only a few common topics (according to state curricular intentions). Such diversity may be desirable, but may well have consequences for assessing aggregate U.S. achievements in science education.
- These plans to cover far more than the average number of mathematics topics, and something closer to the average for science, were true for the grades in which TIMSS is conducting achievement testing** and should make possible some empirical investigation of how effectively we cover such a broad range of topics in comparison to other countries that take a more focused, limited approach to mathematical content, or in which students cover a more extensive common core of topics.

---

*The TIMSS curriculum analysis used hierarchically arranged sets of categories and topics (see Appendix A). School mathematics and sciences were partitioned into sets of subject matter sub-areas here called 'topics.' These sets were described carefully in 'frameworks,' the TIMSS mathematics framework and science framework described in detail in the related, more technical reports. These sets of topics were developed with considerable cross-national feedback and consensus seeking. They were validated by participating country evaluations of field trial curriculum analyses and by linkage to classroom observations and teacher interviews in some countries. The number of topics a country intends to cover may be an artifact of the framework, the list of possible topics. The *comparative* numbers of topics covered by two or more countries is not.

**TIMSS is testing student mathematics and science achievement at three populations. The TIMSS curriculum analysis paid special attention to these populations because of this key link to achievement measures. In particular, textbooks were analyzed as wholes only for the grades in these populations. Population 1 is the two adjacent school grades that contain most 9-year-olds. Population 2 is the two adjacent school grades that contain most 13-year-olds. Population 3 is the final year of secondary school. Important subpopulations of Population 3 are those taking advanced mathematics, advanced physics, or both in the final year of secondary school. In the U.S., Population 1 corresponds to the 3rd and 4th grades, Population 2 to the 7th and 8th grades, and Population 3 to the 12th grade. Since there are some variations in this internationally, when international comparisons are made, we must refer to the appropriate populations. For U.S. curricula we can refer to the appropriate grades.

These findings are discussed in more detail later and their supporting data are presented.

What might these findings imply? Since the U.S. does not devote appreciably more time to science and mathematics education than most countries, U.S. mathematics and science curricula seem clearly to reflect juxtaposed goals competing for the same limited resources of time and attention. Our official mathematics and science curriculum statements have no central focus on a few ideas or approaches to what students should learn in mathematics and the sciences. Whether such a focus is essential to high student attainments is an empirical question. Conventional wisdom and a considerable body of research, suggests that focus and selection are needed when too much is included to be covered well. Focus would seem to be a necessary but not a sufficient condition to high student attainments.

Perhaps we do not need a central focus for our curricula and teaching; perhaps the value of diversity outweighs the value of focus. Perhaps our *de facto* emphasis on breadth will prove more effective overall than other countries' strategies of focusing on strategic topics. That is a matter for further empirical evidence and public discussion. However, it is a clear, empirically documented fact that such central focus has not been typical of recent U.S. mathematics and science education.

Textbooks. Our data show that splintering also shows itself in U.S. science and mathematics textbooks (at least those investigated a few years ago). For example, we found that

- The U.S. mathematics and science textbooks analyzed included far more topics than was typical internationally at all grade levels analyzed.
- Among the U.S. fourth grade mathematics textbooks investigated, the five topics receiving most textbook space accounted for about 60 percent of the textbook on average. Internationally the five most emphasized topics accounted for an average of over 85 percent of textbook space. Among the U.S. eighth grade mathematics textbooks investigated, the five most emphasized topics in non-algebra textbooks accounted for less than 50 percent of textbook space compared to an international average of about 75 percent. In contrast, U.S. eighth grade algebra books were highly focused, the five most emphasized topics accounting for 100 percent of the books.
- Among the U.S. fourth grade science textbooks investigated, the five topics receiving most textbook space accounted for just over 25 percent of the textbook on average. Internationally, the five most emphasized topics accounted for an average of 70 to 75 percent of textbooks. Among the U.S. eighth grade science textbooks investigated, the five most emphasized topics in more general science texts accounted for about 50 percent of textbook space compared to an international average of about 60 percent. In contrast, U.S. eighth grade science books oriented to a single area (physical science, life science, earth science) were highly focused, with the five most emphasized topics accounting for more of the textbooks than was true in the international average.

- The U.S. eighth grade science textbooks investigated emphasized understanding and 'using tools, routine procedures, and science processes.'* This emphasis was typical of what was done internationally. It is not, however, what is typical of the diverse and more demanding performances called for in current U.S. science reform documents. U.S. mathematics textbooks investigated showed a similar emphasis on 'knowing' and on 'using routine procedures' but reflected a more diverse and demanding set of expected performances by students.

- At the time of these investigations, the major mathematics reform document (the NCTM *Standards*) had been actively disseminated for about three years. The current science reform documents (the AAAS *Benchmarks* and the National Academy of Science *Standards*) had yet to be released. These documents emphasize focusing on strategic content and devoting less time to some traditional contents deemed less central. If U.S. mathematics and science textbooks published since the data for this report were collected reflect greater impact of these documents, those books may be more focused and may demand more complex and diverse performances from students. Additional data are needed to determine this. (However, more current data on teacher content coverage suggest that science and mathematics content actually covered in classrooms remains more typical of the earlier textbooks than of the reform documents. Perhaps more time is needed for these reform documents to have their full impact).

Again, these findings are discussed in more detail later and their supporting data are presented.

Our data discussed later make it clear that splintering helps shape the content, even the size, of U.S. science and mathematics textbooks. How might this come to be? Mathematics and science textbooks reflect a cautious, inclusive approach to subject-matter content. Our commercially produced textbooks compete for shares of diverse, often contradictory markets. It is likely that these markets shape the textbooks produced as much as do the ideas of mathematics and science educators. Competing for sales and adoptions in a complex arena, most mathematics and science textbooks opt for a cautious approach. They include something of everything, perhaps to maximize market share.

Debate continues about the circular relations between textbooks and markets. Do markets determine the form and content of textbooks, or does availability in textbooks mostly shape

---

*The TIMSS frameworks include more than sets of content topics. Among other aspects, they include "performance expectations." These are categories of tasks that students may be expected to develop the capacity to perform. These are stated in curriculum plans and underlie each block of a textbook. These performance expectations went through the same cross-national development and validation process as the topics. There are far fewer of them than there are topics in the framework. What is most useful about performance expectations may be not the numbers of expectations included in a document, but the actual expectation and which of a few broad categories it falls into. This helps characterize how diverse and demanding a particular curriculum plan (guideline) or textbook is. Throughout this document, *verbatim* curriculum framework topic or performance expectation labels are indicated with single quotation marks ('numbers,' 'using routine procedures'); informal topic areas (algebra, energy) and references to topics that are part but not all of a formal category are not enclosed in quotation marks.

market demand? The actual process is likely iterative, with the interrelationships not easily resolved. Given this fundamental relatedness and a changing context in school mathematics and science, textbook producers' cautious, inclusive approaches to the diverse goals and intentions of the U.S. in mathematics and science education seem a reasonable strategy. In considering ways to change the *status quo*, however, it may well be that textbook publishers must bear part of the risk of producing change.

Instruction. Splintering also affects instruction. We found among other results, for example, that

- U.S. eighth grade mathematics and science teachers typically teach far more topic areas* than their counterparts in Germany and Japan. (The full TIMSS cross-national data set for teachers has not yet been released for discussion.) This is true for science teachers even when using a single area textbook (physical science, life science, earth science).
- U.S. eighth grade mathematics teachers surveyed indicated that they taught at least a few class periods on almost every topic area included in the survey's questionnaire. In contrast, they devoted 20 or more periods of eighth grade instruction to only one topic area, fractions and decimals. Other additional topic areas received this more extensive coverage in Germany and Japan.
- U.S. eighth grade *general* science teachers also indicated that they would devote at least some class time to every topic area surveyed. None was omitted completely. None was marked to receive more than 13 class periods of attention by eighth grade physical and general science teachers. Additional topic areas received more extensive coverage in Germany and Japan.
- The five surveyed topic areas covered most extensively by U.S. eighth grade mathematics teachers accounted for less than half of their year's instructional periods. In contrast, the five most extensively covered Japanese eighth grade topic areas accounted for almost 75 percent of their year's instructional periods. On average, U.S. eighth grade general science teachers' most extensively covered topics accounted for only about 40 percent of their instructional periods, but this percentage was also lower for science in Germany and Japan (about 50 to 60 percent).
- More than 75 percent of U.S. mathematics teachers indicated familiarity with the NCTM *Standards*. Fewer U.S. science teachers indicated similar familiarity with the corresponding science reform report, but it was more recent than the mathematics report

*To survey what content teachers covered, the topics from the appropriate TIMSS framework, mathematics or science, were grouped into a smaller number of grade-appropriate categories easily recognizable by teachers. Topics less appropriate to a grade level whose teachers were surveyed were collapsed into fewer, more general categories. The result was still a diverse array of contents for which teachers were asked to indicate roughly how many periods they would devote to coverage during the school year.

by 5 years. These data suggest, however, that, given enough time, failure to see widespread change in teacher practice may not be simply due to a lack of information.

- U.S. science and mathematics teachers are scheduled for far more class periods per week than their counterparts in Germany and Japan (and, presumably, in other countries as will become clear when the full teacher data set is released for reporting). This suggests that the lack of desired changes or effectiveness is not due to lack of effort on teachers' parts.

These findings (and their counterparts for other grades) suggest that while at times demanding, content coverage by U.S. teachers spreads its attention among many topics and only occasionally focuses comparatively more attention on a few topics. However, U.S. teachers do use a somewhat more diverse and demanding set of activities with students. As before, these findings are discussed and their supporting data presented in more detail later.

How could this unfocused pattern of teacher content coverage come about? Mathematics and science teachers rarely have the luxury of being idealists. They live in real worlds of crowded classrooms, competing demands, ambiguous educational goals, broadly inclusive textbooks, and shifts in approved practices. We should not be surprised that the classroom practices of teachers found in this situation also reflect the cautious, inclusive approach of textbooks. Teachers, too, often cover something of everything, and little of any one thing. They use a variety of classroom activities and approaches. We may, perhaps, find it surprising (as our data show) that the beliefs and pedagogical preferences of U.S. mathematics and science teachers are so much more adaptive than their actual practices.

Reform. The U.S. has the benefit of some of the best advice available on recommended reforms in science and mathematics education. However our findings suggest

- State mathematics and science curriculum guides, plans, and statements of intentions as late as two to three years ago still called for coverage of far more topics than most other countries did and, far more than would be indicated by current reform agendas in mathematics and science education. This does not imply that reform efforts have been futile, but, at most, this implies that reform may take considerable time (since efforts at reform were comparatively new at that time). Whether the reform agendas' recommendations to focus on more strategic content and downplay other content have been heeded in more recent state curricula requires additional data gathering using more recent documents.

- Mathematics and science textbooks as recent as two to three years ago took an inclusive approach that included some reformed emphases in mathematics (the major science reform documents not yet having been released) but did not drop more traditional contents. The results were books that devoted space to many topics and focused little on any particular topic. This, again, is contrary to the recommendations of current reform, and additional data collection is needed to verify whether more time has provided additional impact from reform documents.

- Fairly recent data on teacher instructional activities indicate that teachers continue to take an inclusive approach to content coverage but that they do use diverse and, at times, powerful activities and expectations with students. The teachers indicate some familiarity with reform documents. Their practice, however, says that the impact of these reform recommendations may be less than what might have been hoped for.

The most obvious explanation for the limited impact (at the time of the majority of our investigations) of reform recommendations is that more time is needed for widespread change. Our national samples could only reflect widespread, common change by many teachers and not more limited changes — despite the fact that limited changes might be important as focal points for eventual more wide-spread change.

Other explanations may be needed as well. In the U.S. today, we live in a climate of reform and discussed reform. Professional organizations concerned with mathematics and science education issue platform documents setting out agendas, benchmarks, and "standards" for changed policies, curricula, teaching practices, and desired student attainments. These calls for reform are powerful, demanding, insightful, and offer coherent visions of what might be done to make major improvements in their targeted educational practices. They help to articulate focused, strategic content — both in what they recommend receive more emphasis and what they recommend receive less. They portray diverse and powerful expectations for students. However, our data indicate that, until a few years ago, the impact on state curricula and textbooks was limited and, more recently, the impacts on teaching practices were hard to identify. Why might that be?

Our "national" vision of what to do in mathematics and science education takes place in thousands of sites — local, state, and federal. It is a conversation of many voices — those responsible for educational decision-making, textbook and test publishers, members of professional organizations, and others. Reform documents themselves often emerge from compromise among professionals, and this compromise may strain their pursuit of coherent visions. In this context of a composite vision, urgings for reform, no matter how intellectually coherent they are in themselves, unfortunately may become part of a babel of voices stating competing, at times conflicting, demands for practice, change, and reform in science and mathematics education. Although these reform agendas may have coherent and powerful rationales, they must be interpreted, adopted, and implemented in thousands of locations to bring widespread change. It is perhaps unsurprising that our data show that the actual effects of recommended reforms in mathematics and science education have often reflected much less impact than might be wished.

Given what the data reveal about how reforms actually enter our distributed "national" educational system, slow progress is certainly no reflection on the quality or power of mathematics and science reform efforts that have yet to be as effectual as they wish in this climate. Certainly it would be drastically wrong to conclude, on the basis of progress thus far — especially as indicated in the "early returns" of our data — that these reforms have failed. Rather, it seems more appropriate to be amazed at their current successes. Clearly our system of shared educational responsibility and past fragmented practice set boundaries that have limited these educational reforms and may continue to do so. If we seriously believe that we need widespread change, we must seek to understand our mathematics and science education practices better and to identify structural changes that may facilitate well-conceived efforts at content-based reform. Only by doing so are we likely to maximize the chance for effective mathematics and science education reform in our national context.

Expected student attainments. What should we expect from a composite educational system with a long tradition of unfocused, inclusive practice in science and mathematics education that is yielding, at best, slowly to reform efforts? The expectations differ for science and mathematics education; although, there are some similarities. Our data suggest that

- The U.S. composite mathematics curricula included virtually all of the international composite.
- The U.S. composite mathematics curricula contained virtually all of the international composite *and more*. Unfortunately, that "and more" reflects the typical unfocused, inclusive approach of many U.S. curricula, textbooks, and teachers. While it remains an empirical question whether this lack of focus will prove detrimental to comparative educational achievement in mathematics, the conventional wisdom predicts that it will. The U.S. composite science curricula was comparatively more focused but often in ways in which only some students received a focus on a particular area of science (physical, life, or earth sciences).
- U.S. Algebra I classes were far more focused, and better achievements for them can be expected in the areas on which they focus, and perhaps in other areas as well.
- What was "basic" in mathematics in the U.S., Japan, and Germany differed. The U.S. mathematics instructional practices defined *de facto* basics of arithmetic, fractions, and a relatively small amount of algebra. The impact of this on a test reflecting a cross-national test of the mathematics "basics" remains to be seen.
- What was "basic" in science corresponded more with the international core than was true for third and fourth grade mathematics. For eighth grade, the picture is more complex since considerable use was made of single area courses (physical sciences, life sciences, or earth sciences). These courses defined a more restricted, focused set of basics,

but they applied only to the subset of students receiving those particular courses. The effect on aggregate achievement is difficult to predict.

- Clusterings of TIMSS countries based on profiles of content areas show that the U.S. is not often clustered with those countries with which it might most naturally seek comparisons — the G-7 countries, the European Union, etc. While economic competitiveness is not the sole reason for education in mathematics and the sciences, to the extent it is an important reason, we have cause for concern if achievement is related closely to content differences. The answer to this empirical question must await the release of the TIMSS achievement data sets.

Certainly, in view of these findings it seems wishful optimism to expect our students to achieve highly in science and, especially, mathematics compared to students in other countries. Whatever other benefits our fragmented system may have, it seems likely to limit our students' successes.

Most nations do not share similarly splintered visions in mathematics and science education. Many of their practices reflect more coherent visions. While central guiding visions do not alone guarantee student achievement, they seem often to contribute to optimal attainments. That is, these shared visions are insufficient to ensure desired achievements, but they seem necessary. TIMSS achievement data analyzed in the context of these curricular analyses can help to reveal whether this conjecture is true. Many other nations' student achievements undoubtedly will reflect the power of coherent goals and firmly grounded teaching practices. Those nations whose educational systems do share our pattern of splintered visions — our true "peers" in mathematics and science education practices — will not often be those with whom we wish to be most directly compared.

Chapter 1

UNFOCUSED CURRICULA

This section focuses on intended, planned curricula and presents findings and supporting data in more detail. We begin with the idea that there are composite U.S. curricula in mathematics and in the sciences (although the two differ greatly). Just as the products of internationally competitive U.S. corporations help determine U.S. international competitiveness, something creates the yield of U.S. education in the sciences and mathematics. That "something" we call a composite curriculum. It is not a U.S. national curriculum, that is, an official, federally mandated document or policy of what must be taught in mathematics and the sciences. It is the aggregate policies, intentions, and goals of the many educational subsystems making up the loose U.S. federation of guiding educational visions in the sciences and mathematics.

A particular city or state provides educational experiences in the sciences or mathematics. Those activities and their results not only are educational activities of that city or state but also are part of a composite U.S. curricular experience. Student achievements flowing from these curricular experiences are part of U.S. educational yield, as measured in national achievement tests in mathematics and the sciences.

This composite curricular experience is fundamentally splintered, unsurprising for a union of loosely related activities. Are there common features among the subsystems sufficiently strong for U.S. to identify characteristic features of these curricular efforts? The TIMSS curriculum analysis was based primarily on state curriculum frameworks or guides and on supporting opinion by experts in mathematics and science education. We drew an appropriate random sample of state curriculum guides in 1992-93*. We then carefully aggregated them statistically to portray the current composite mathematics and science curricular intentions of the states. Any marked features of the composite reveal widely common patterns in the states.

---

*Selecting documents for the U.S. Curriculum Analysis presented considerable challenges given the nature of curriculum policy making and textbook markets in this country.

Regarding Curriculum Guides: The decision was made to draw separate stratified samples with probabilities proportional to

How Many Topics Do We Plan to Cover?

What are the dominant characteristics of these composite curricula? First, the composites — and virtually every state document sampled — intended to cover many topics. The planned coverage included so many topics that we cannot find a single, or even a few, major topics at any grade that are the focus of these curricular intentions. These official documents, individually or as a composite, are unfocused. They express policies, goals, and intended content coverage in mathematics and the sciences with little emphasis on particular, strategic topics.

The U.S. consistently intends to cover far more than the typical number of topics (see Exhibit 1). All countries plan to cover comparatively more topics in the middle grades and to focus in later grades. Japan intends to begin to focus attention on fewer topics in grade 6. Germany plans to cover fewer topics generally and especially in grades 9 to 12. The U.S. likely does not focus its mathematics curriculum plans on a few strategic topics since so many topics are planned to receive attention in each grade. Other technical reports of the TIMSS curriculum analysis[7] give detailed descriptions of the methods that produced these and other curriculum analysis data.

Similar findings are true for science education (see Exhibit 2). The U.S. propensity to plan to cover more topics is not as pronounced for science as for mathematics. With the exception of specialized secondary school courses, we think it unlikely that state science curricula can focus on strategic topics when they plan to cover so many. The U.S. intends to cover more science topics than Germany and Japan in the early grades and to specialize in the upper grades. Germany intends to cover fewer topics than the U.S. at the early and later grades — specializing a year earlier than the U.S. Japan, on the other hand, intends to cover fewer topics than the U.S. at all but grades 11 and 12. This suggests that Japan follows an integrated, multidisciplinary approach throughout secondary school, rather than organizing science into a few single-discipline courses. Alternatively, Japanese students may typically be meant to cover more than one single-discipline science course.

---

size (PPS) of 15 state curriculum frameworks. The criterion used for the PPS sampling was the school-age population of each state. The strata considered for the sample (drawn in 1993) were:

1. Recency: Whether or not guides were published before or after 1988.

2. Impact: Whether or not the curriculum framework was used in state assessment or in state textbook adoption schemes.

3. Region: Sample was drawn to ensure representation of five regions: New England/Mid-Atlantic, South, Midwest, Southwest, and West.

Regarding Textbooks: Selection of textbooks was especially problematic given the nature of the textbook market in the U.S. Publishers refuse to share data on sales, and no other recent data on textbook use in the U.S. were available for drawing a sample. The goal was to obtain a sample of textbooks that together represented those textbooks used by at least 50 percent of the national school population in grades 4 and 8 as well as the textbook that was most representative of students in advanced mathematics and physics in grade 12. Given this, experts were surveyed and lists of textbooks adopted in large states were used to draw the final sample. For Populations 1 and 2 mathematics and science, the lists of textbooks were rank ordered from those textbooks most commonly used to those least commonly used. Percentages of students using these books were calculated. Starting from the top of the list, textbooks were chosen until the percent of students represented reached approximately 50 percent. For Population 3, the mathematics and science book with the largest percentage of student use was selected.

Final Sample: These procedures resulted in the U.S. sample of state curriculum guides and textbooks listed in Appendix B.

Exhibit 1.

Number of mathematics topics intended.

On average, the states indicated plans to cover so many topics that the U.S. composite shows an intention to cover more mathematics topics than the majority of other countries. The number of topics to be covered dropped below the 75th percentile internationally only in grades 9, 10, 11 and 12 when we typically teach mathematics in specific courses - algebra, geometry, etc. [The gray bars show how many mathematics topics were intended to be covered at each grade in the TIMSS countries. The bars extend from the 25th percentile to the 75th percentile among countries. The black line indicates the median number of topics at each grade. We marked the U.S., Germany, and Japan individually.]

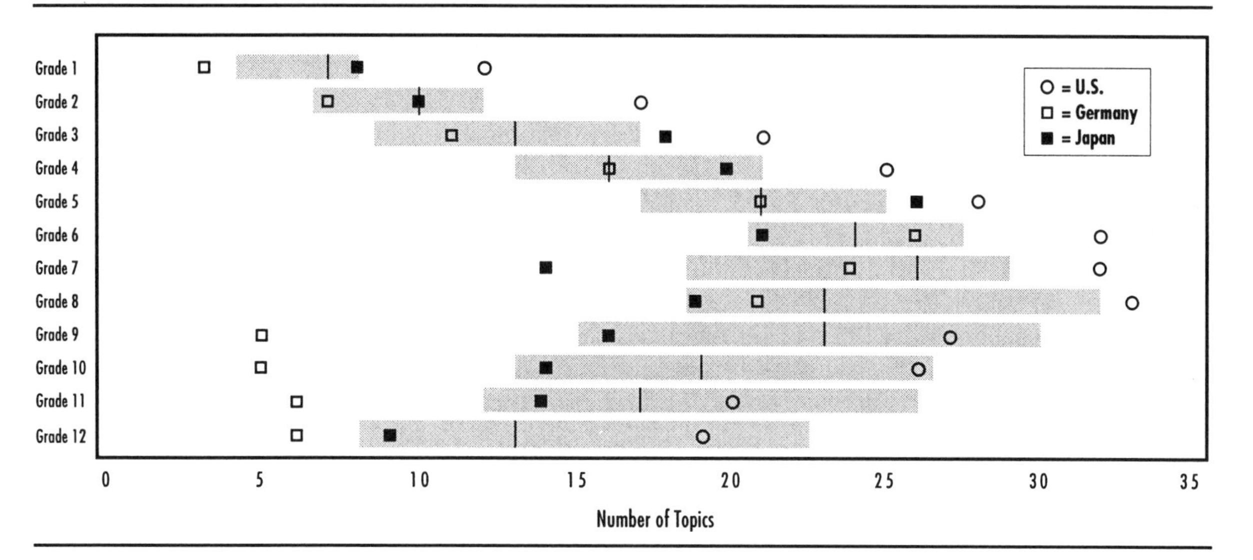

Exhibit 2.

Number of science topics intended.

On average, the states indicated plans to cover so many topics that the U.S. composite shows more science topics generally covered than the median for other countries. The number of topics to be covered dropped significantly only in grades 10, 11, and 12 (below the 25th percentile in grades 11 and 12). We appear, in these later grades, to have abandoned more general sciences approaches for specific courses — chemistry, physics, etc. [The gray bars show how many science topics were intended to be covered at each grade in the TIMSS countries. The bars extend from the 25th percentile to the 75th percentile among countries. The black line indicates the median number of topics at each grade. We marked the U.S., Germany, and Japan individually.]

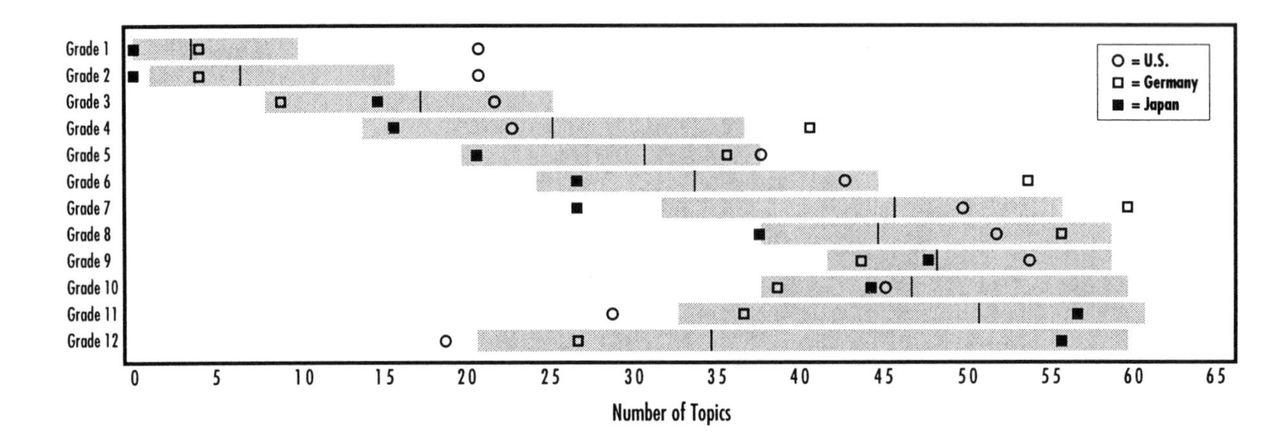

The U.S. intends to cover comparatively more topics than most other countries in mathematics and (less so) in science. What is behind this pattern? Do we typically fragment mathematics and science content into more (and, presumably, smaller) content topics than other countries? If our topics are less inclusive and we intend to cover more topics overall, then we would need to cover more topics at each grade.

If so, why do we fragment mathematics into more pieces and force ourselves to cover more of these smaller pieces in each grade? In the U.S., school mathematics has a long history of fragmenting content and gradually building up central topics. Traditionally, mathematics concepts and skills have been considered difficult in "large pieces." However, when broken into smaller facts and subskills, these smaller pieces are more easily mastered. They can be built up and accumulated by students gradually. Thus, our typical practice in mathematics is to accumulate facts and skills incrementally. In theory, if not always in fact, we occasionally provide educational experiences designed to integrate these smaller facts and skills into larger concepts and more complex skills. This conventional practice may work for simpler facts and routine skills. It certainly seems less appropriate for the more complex, integrated mathematical expectations recommended in current mathematics reform proposals.

Science curricula are also fragmented. However, this fragmentation likely has a different basis than that for mathematics. School science, unlike school mathematics, deals with content from several sciences. Dividing curricular attention among many small topics may make it easier to cover at least some of several different science areas. However, the U.S. composite science curriculum has a relatively small number of core topics common across many states. We do not cover more topics in our *composite* science curriculum because science topic choice varies more among the *states*. The sampled states consistently planned to cover more topics than other countries did, but they were not the same topics from state to state. Even the U.S. composite science curricula at various levels typically included as many or more topics as the international composite. The result was a smaller common core of topics in the U.S. composite.

Science education was understood differently in individual states resulting in each state choosing many particular topics that were different from those in other states. We can only speculate on the conceptualization of science education that would allow such inter-state variety in the choice of topics to be taught.

Do we typically leave individual topics in our curricular plans for more grades than do other countries? This would be consistent with a philosophy that continued attention and revisiting topics over the grades leads to greater mastery. However, it neglects the advantages of focused attention. The alternatives seem to be to "distribute mastery" over the grades with cumulative learning, or to focus more intensely on a few strategic topics at any one time and achieve mastery through intensive attention. Certainly, the U.S. composite curricula as measured were comparatively unfocused. Achievement scores may help us know whether distributed mastery and cumulative learning yield comparatively more than focused, intensive coverage.

We can answer our questions of whether and why we fragment content into more topics than do other countries only by further investigation. The TIMSS data have other analytic possibilities for answers not yet explored. The current data can tell us whether we plan continued topic coverage for more grades than do other countries. We should also remember that these two possibilities for creating more topics are not mutually exclusive. We may both fragment content into more, smaller topics *and* leave each topic in our curricula for more grades than do other countries.

Do We Plan to Add More Topics Than We Drop Over the Grades?

Do we accumulate so many composite curriculum topics because we typically add more topics than we drop from intended curriculum coverage? Alternatively, do we accumulate so many topics because we add but rarely (if ever) drop topics? The first pattern implies we continually change the topics we attend to in our states' curricula, shuffling topics in and out over the years. It also suggests that we rarely drop as many as are added thus creating a residual accumulation of topics that continue to be covered. In this pattern, we would change content but perhaps *unintentionally* accumulate topics through continued review, failure to set mastery points for specific contents, etc. The second pattern implies that we *intentionally* leave the same topics in our curricula for long periods; that is, it is a pattern of distributed mastery with the same topics intentionally continued for several grades.

We commonly plan to add far more mathematics topics than other countries do in grades 1 and 2 (see Exhibit 3). We commonly plan to drop no topics until grade 7. So those early topics and those introduced afterward remain for several grades. We do not plan to add more than is typical in most grades after grade 3 except at grade 9. Our pattern is to introduce topics early and have them persist into secondary school.

Grade 9 is special in the U.S. composite mathematics curriculum. We are at the 75th percentile in the number of topics planned to be added, and we indicate plans to drop far, far more than even the 75th percentile of other countries. This clearly indicates common plans to change from a yearly coverage of many topics to coverage of fewer, more specialized topics. This grade, of course, historically has corresponded to major enrollments in Algebra I in most states of the U.S. and apparently still does. A similar pattern of dropping many topics and adding only one is typical for grade 11. This again suggests a more focused, limited-content course, often Algebra II.

There are very few changes in grade 12 mathematics. This is likely an artifact of previous specialization to algebra in grade 11 and of most topics of interest having been previously introduced. Grade 10 is more puzzling. The common intentions are that no new topics be introduced and only a few dropped. Perhaps this reflects plans to introduce earlier all categories of geom-

Exhibit 3.

Number of mathematics topics added and dropped.

After grade 2, the composite U.S. mathematics curriculum does not indicate an intention to add comparatively more topics. The U.S. is within the middle 50 percent of TIMSS countries and at or below the median in almost every case. However, the U.S. does not show any topics as dropped until grade 7 and drops many more than other countries only in grades 9 and 11. Topics introduced in the early grades remain until junior high school and comparatively large numbers of topics are dropped only with specialization in high school. [The gray bars on the right show how many mathematics topics were intended to be added at each grade in the TIMSS countries. The bars extend from the 25th percentile (nearest the center line) to the 75th percentile among countries in the number of topics at each grade. The gray bars on the left show how many mathematics topics were intended to be dropped from coverage at each grade in the TIMSS countries. The bars extend from the 25th percentile (nearest the center line) to the 75th percentile among countries in the number of topics at each grade. The black line at each grade indicates the median number of topics. The U.S., Germany, and Japan are marked individually. Missing symbols for the U.S., Germany and Japan indicate no topics for the category (added, dropped, or both).]

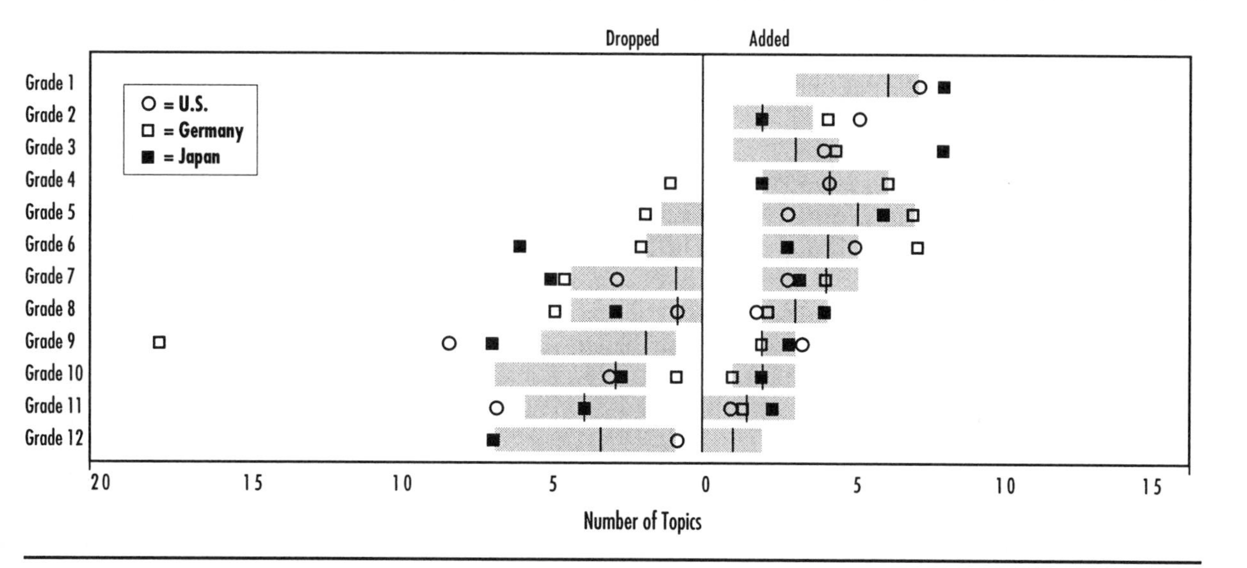

etry and algebra examined by the TIMSS analysis. The data do not rule out a specialized course in the U.S. at grade 10. This course, however, would consist almost entirely of renewed (perhaps more formal) coverage of topics previously introduced.

The U.S. composite science curriculum is structured very differently (see Exhibit 4). We commonly plan to drop no science topics before grade 10. This appears to be our typical time for shifting to single-discipline, specialized courses such as chemistry, physics, etc. Prior to grade 10, we plan to add many, many more topics at grade 1 than do other countries. These remain for several grades. In grade 5, we again intend to add comparatively many topics, still without dropping any. A few topics are planned to be added in grades 6 to 8. Since no topics are planned to be dropped, the core of topics introduced in grade 1 commonly persists and is supplemented by a large and slowly growing number of topics through grade 9.

This suggests a multidiscipline, general science approach to science education through grade 5 with topics revisited and explored further. In grades 10 to 12 we commonly plan to drop unusually many topics with virtually no additions. This is characteristic of specializing the previous science content into more focused, probably single-discipline courses.

We can "track" individually only Germany and Japan. Germany's pattern is roughly like the U.S. but with little science taught until grade 4. Japan's pattern is very different. Japan seems to intend dropping very few topics other than in grades 10 and 11. It plans to add unusually high numbers of topics periodically (first in grade 3, when science education begins, and then in grades 8, 9 and 11). This suggests a more integrated, general, or extensive pattern of science content coverage even in high school. If specialization occurs in Japanese high school science education, it seems likely to have a very different basis than the single-discipline, focused courses of the U.S. and Germany.

As we said earlier, there is clear evidence that we introduce many topics and intend them to persist over several grades in mathematics and, to a lesser extent, in science. The common feature appears to be that of persistent topics, either due to some philosophy of distributed mastery (mathematics) or some general science approach that involves exposure to unusually many topics (science). In both cases, we must characterize the curriculum as unfocused, except in a few specialized high school courses.

How Long Do We Plan Continued Study of a Topic?

Our discussion so far suggests that we plan for many topics to "come early and stay late" in the U.S. composite mathematics and science curricula. The data further reveal some broad categories of the particular topics in which the U.S. is atypical. To reveal this we look at the *duration* of a topic — how many grades we plan for it to remain in the composite curriculum after its introduction.

Exhibit 4.

Number of science topics added and dropped.

We seem to plan introducing new science topics at intervals - especially grade 1 and grade 5, with little change in the intervening grades. We also seem commonly to plan dropping many topics in grades 10-12 while adding few or none. This is characteristic of a shift from an integrated, "general science" approach to single-discipline courses in high school. [The gray bars on the right show how many science topics were intended to be added to coverage at each grade in the TIMSS countries. The bars extend from the 25th percentile (nearest the center line) to the 75th percentile among countries in the number of topics at each grade. The gray bars on the left show how many science topics were intended to be dropped from coverage at each grade in the TIMSS countries. The bars extend from the 25th percentile (nearest the center line) to the 75th percentile among countries in the number of topics at each grade. The black line at each grade indicates the median number of topics. We mark the U.S., Germany, and Japan individually. Missing symbols for the U.S., Germany and Japan indicate no topics for the category (added, dropped, or both).]

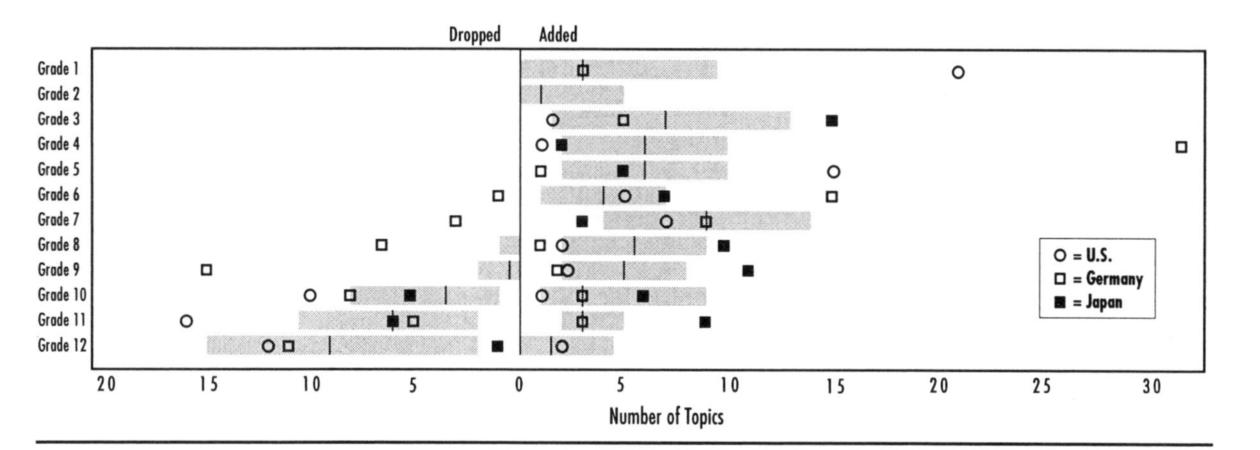

Do the U.S. composite curriculum topics on average have longer planned durations than those of other countries? This is somewhat true for science (see Exhibit 5). The average number of grades that science topics are intended to remain in the U.S. composite curriculum is about two grades greater than the international median. Only six countries have higher average planned durations. Many individual U.S. science topics, as well as this average over all science topics, are planned to persist longer than for most other TIMSS countries.

The average number of grades that mathematics topics are intended to remain in the U.S. composite curriculum is also about two grades greater than the international median (see Exhibit 6). In mathematics, only five countries have higher average planned durations. For mathematics or science, we commonly intend for individual topics to remain in our curricula longer than do most other countries. This is further evidence that U.S. mathematics and science curricula typically intended to deal with topics for more years as well as to deal with more topics. A pressing question is whether having topics "come early and stay late" in our science and mathematics curricula is correlated with planning to cover "a little of everything but not much of anything." Given that each year has limited time allocated for mathematics and science education, this correlation seems likely.

Are there particular mathematics content areas that force the average duration of topics in the composite curriculum higher than is typical? The TIMSS curriculum analysis was in terms of hierarchical topic sets in which more specific, limited topics were grouped as parts of larger, more general categories. This allows us to analyze average durations for all specific topics within a more general category of mathematics or science topics — for example, 'numbers' or 'life sciences.' The results more clearly show what content areas typically persist in U.S. curricula.

The U.S. average intended durations are about two grades higher than the international median for 'numbers' and 'measurement' topics. The U.S. is fairly close to the median for equations-related algebra, not because it does not spend several years on this topic but because virtually every other country does as well. Most of the other mathematics topics considered (excluding advanced mathematics — calculus, probability, etc.) are parts of geometry, 'proportionality,' function-related algebra, and 'data representation and analysis.' This combined set of topics explain much of the way that the U.S. differs from other countries in average intended duration of mathematics topics. As Exhibit 7 shows, the U.S. is over two grades higher than the median for the topics in this aggregate category, higher than all but three other TIMSS countries.

Thus, our data indicate we plan coverage in more grades than is typical for 'numbers' and 'measurement' topics, and even more so for geometry, proportionality, functions, and data representation. From this we see the U.S. does not have a completely arithmetic-driven curriculum persisting on into junior high school. We have planned time for a variety of nonarithmetic topics (as well as many arithmetic topics). However, we seem to get this additional topic coverage

Exhibit 5.

Average intended science topic duration.

The average number of grades science topics are intended to remain in the U.S. composite curriculum is well above the international median and greater than all but six other TIMSS countries. The average durations for Germany and Japan are at or below the median. [These data are the intended duration of science topics averaged over all science topics for each country. The intended duration is centered on the median for all countries. Those appearing to the right have greater than typical average science topic duration.]

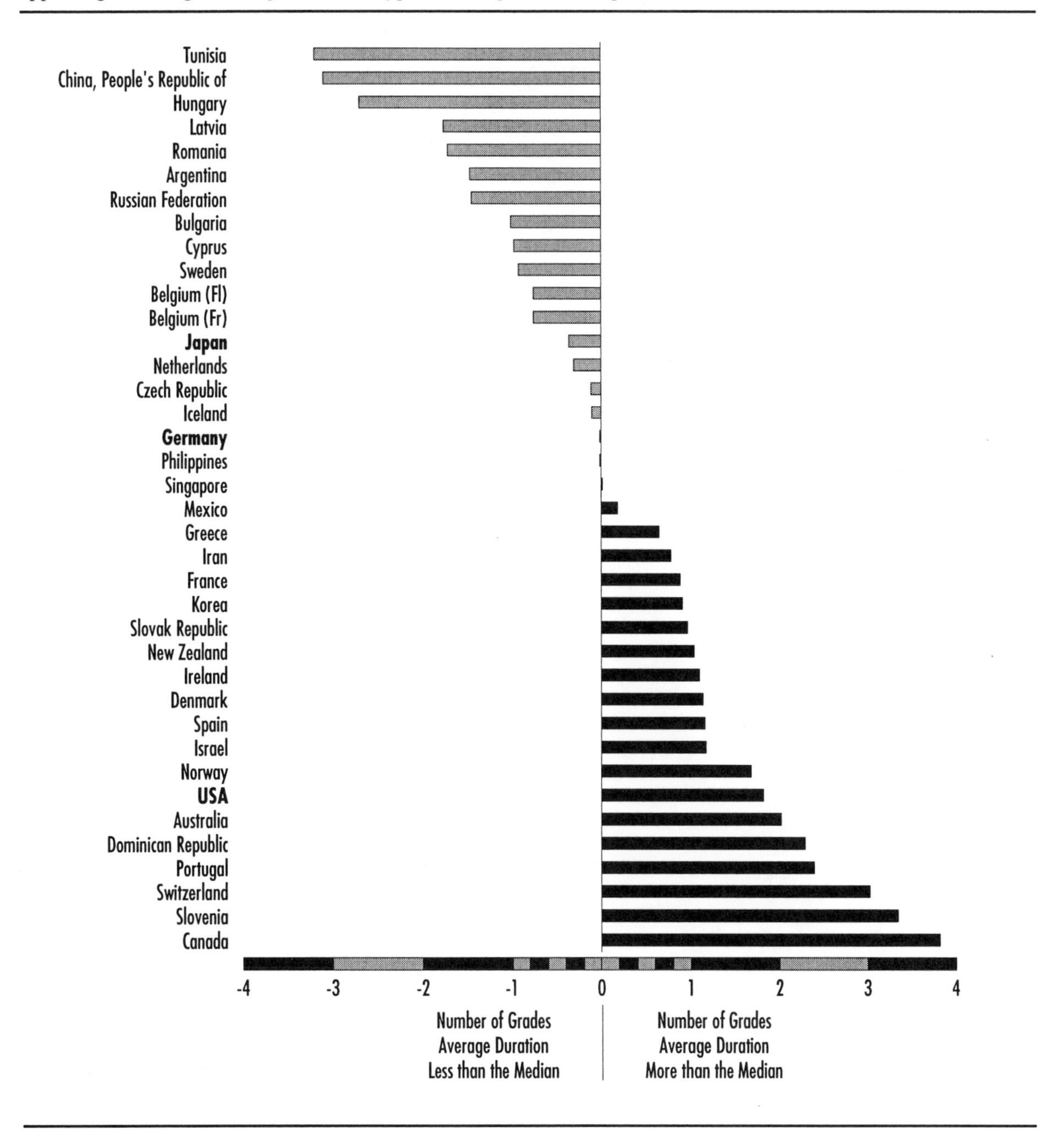

Exhibit 6.

Average intended mathematics topic duration.

The average number of grades that mathematics topics are intended to remain in the U.S. composite curriculum is also well above the international median and higher than all but five other TIMSS countries. The average durations for Germany and Japan are below the median, Germany markedly so. [These data are the intended duration of mathematics topics averaged over all mathematics topics for each country. The intended duration is centered on the median for all countries. Those appearing to the right have greater than typical average mathematics topic duration.]

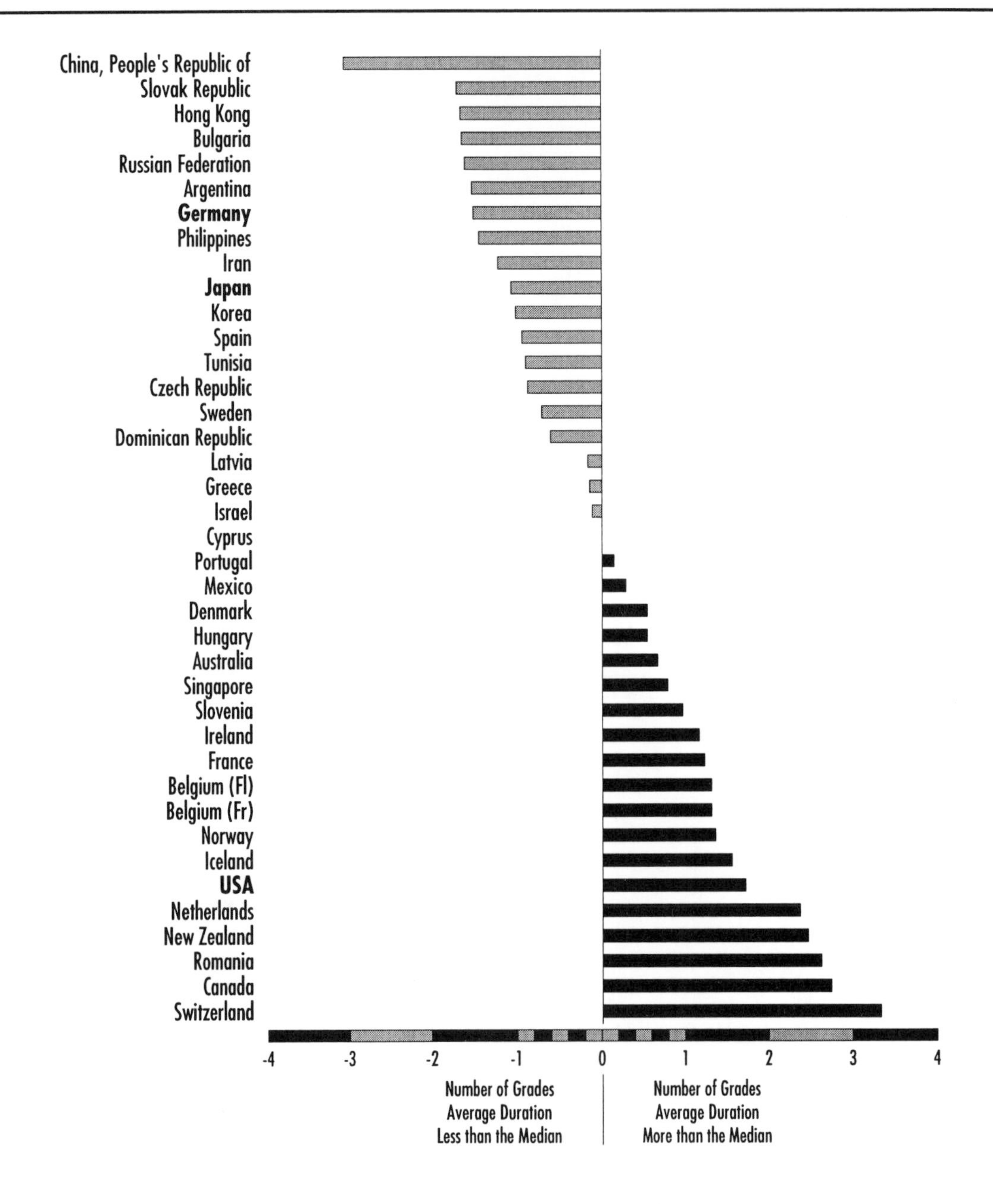

Exhibit 7.

Average intended duration for selected mathematics topics.

The average intended durations of geometry, proportionality, functions, and data representation topics in the U.S. are well above the international median duration. They are higher than all but three other TIMSS countries for topics in these categories. Japan is about at the median, but Germany markedly below the median. [These data are the intended duration of mathematics topics in geometry, proportionality, functions, and data representation, averaged over all topics in these categories for each country. The intended duration is centered on the median for all countries. Those appearing to the right have greater than typical average mathematics topic duration.]

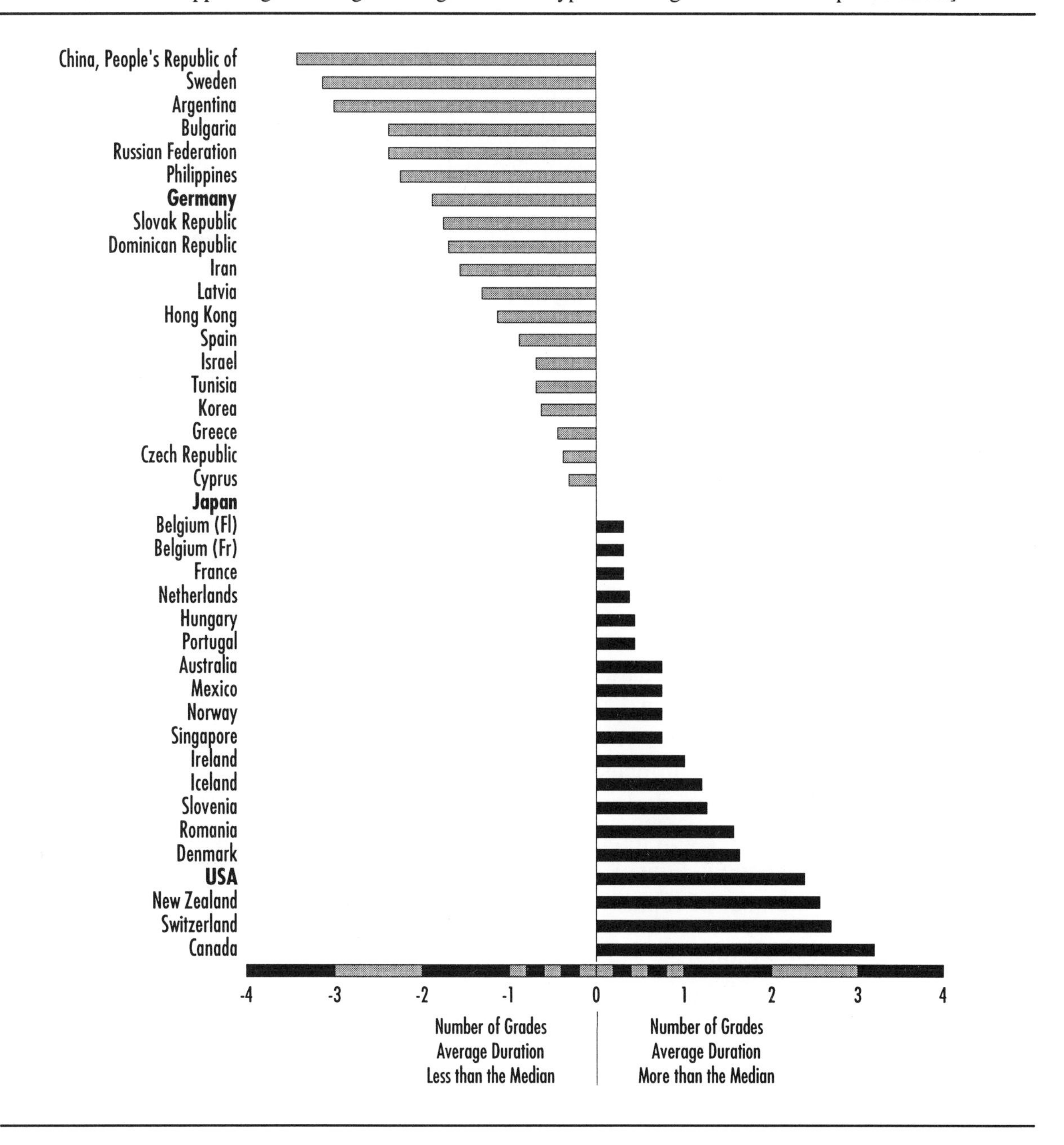

at the price of letting topics persist longer and of devoting attention to more topics than usual for each grade. We have, perhaps, succeeded in "modernizing" our state mathematics curricula to include a wider range of mathematics, but we have done so at the price of depth and focus. For the best of intentions, we may have created in our states commonly unfocused planned mathematics curricula. It remains, of course, an empirical question of how strongly curricular focus is related to achievement.

Is there a counterpart that drives the average duration for science topics? We classify science topics most broadly into earth science, life science, and physical science topics (including both physics and chemistry contents). The averages across the topics in 'earth sciences' and 'life sciences' are close to the international median. Within the 'physical sciences' (Exhibit 8), the U.S. average is about two grades above the median, higher than all but seven countries. Average intended durations are not as unusual for science as for mathematics. The planned time devoted to physical science topics explains much of what is unusual for the U.S. This does not contradict the earlier claim that U.S. science education focused on more topics longer. It reflects, instead, that many other countries also had comparatively unfocused science curricula. The U.S., as the others and somewhat more so, had an unfocused curriculum in science as well as in mathematics.

How Many Topics Do We Plan to Cover for the Student Populations TIMSS is Testing?

Practicalities in cross-national comparative achievement testing limited TIMSS testing to three populations. Focusing on the upper grades of Populations 1 and 2 and on the specialists in Population 3 tells us more directly whether more topics were planned to be covered at these three points in the composite curriculum. This focus on key grades allows detailed analyses of the sampled state curriculum documents to determine whether intending to cover more topics holds widely among the states.

The U.S. covered more mathematics topics for the populations to be tested with one exception, the upper grade of Population 1 (see Exhibit 9). This exception was below the 75th percentile but well above the median of other countries. If covering more topics — and its presumed concomitant of having unfocussed curricula commonly across the states — relates to achievement, we should notice these effects in TIMSS achievement testing. This should be especially so for Population 2 and the mathematics specialists of Population 3. The numbers of science topics were far more typical except in Population 3 (see Exhibit 10). Achievement effects based on the number of topics are likely only for physics specialists.

We should compare Exhibit 9 with Exhibit 1 and compare Exhibit 10 with Exhibit 2. The data in Exhibit 9 and Exhibit 10 are more representative of what we found in any single state curriculum guide. The pattern of covering more topics is present but not as pronounced as in the composite curricula portrayed in Exhibit 1 and Exhibit 2, at least at these selected grades and

Exhibit 8.

Average intended duration for physical science topics.

The average number of grades physical science topics are intended to remain in the U.S. composite curriculum is above the international median and higher than all but seven other TIMSS countries for topics in these categories. Japan is also above the median. Germany is at the median. [These data are the intended duration of science topics in physics, averaged over all topics in these categories for each country. The intended duration is centered on the median for all countries. Those appearing to the right have greater than typical average science topic duration.]

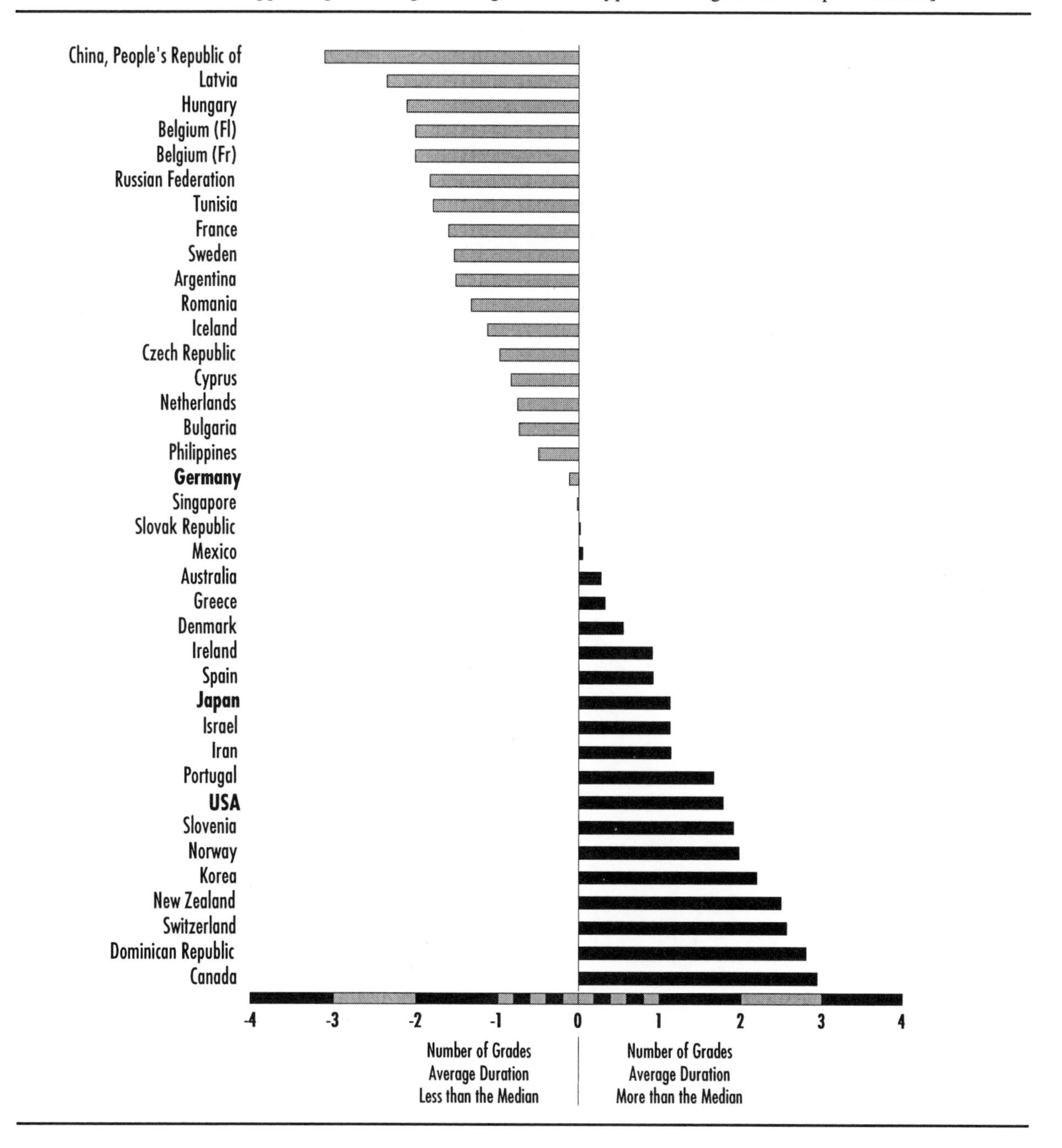

Exhibit 9.

Number of mathematics topics intended at the TIMSS tested populations.

The average of U.S. state mathematics curriculum guides — for the upper grades of Populations 1 and 2 and the mathematics specialists in Population 3 — indicate the states widely planned to cover more topics than most other countries. Only in Population 1 did the number of mathematics topics fall below the 75th percentile of those topics planned for coverage by other countries. [The gray bars show how many mathematics topics were intended to be covered at the corresponding subpopulations in the TIMSS countries. The bars extend from the 25th percentile to the 75th percentile of the number of topics among countries at each population. The black line indicates the median number of topics at each population. We marked the U.S., Germany, and Japan individually.]

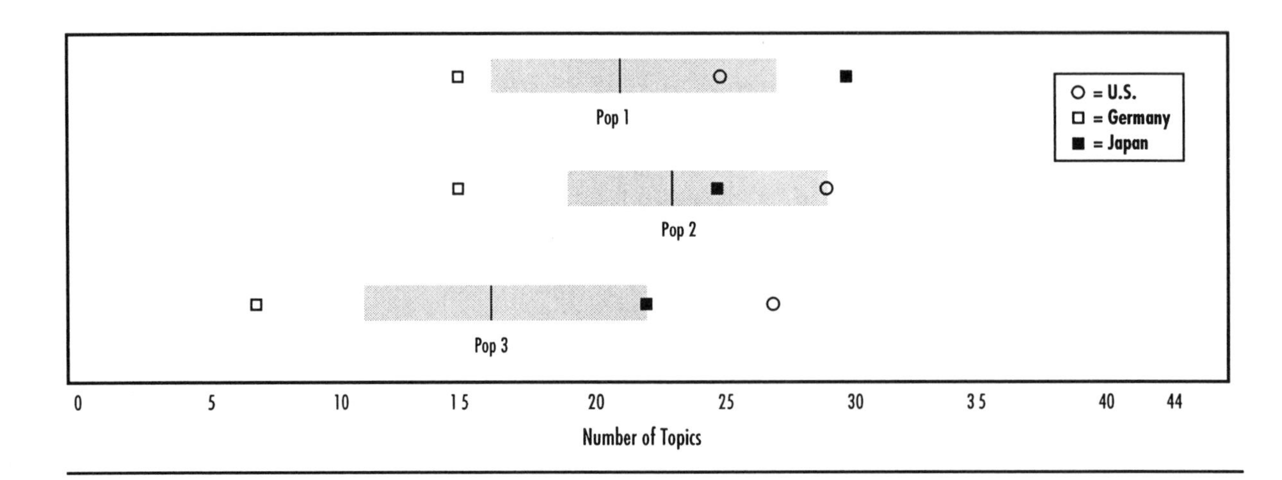

Exhibit 10.

Number of science topics intended at the TIMSS tested populations.

The average of U.S. state science curriculum guides — for the upper grades of Populations 1 and 2 and the physics specialists in Population 3 — show the states planned to cover fairly typical numbers of topics, except in the case of Population 3. [The gray bars show how many science topics were intended to be covered at the corresponding subpopulations in the TIMSS countries. The bars extend from the 25th percentile to the 75th percentile of the number of topics among countries at each population. The black line indicates the median number of topics at each population. We marked the U.S., Germany, and Japan individually. German curriculum guide data were not available for Population 1.]

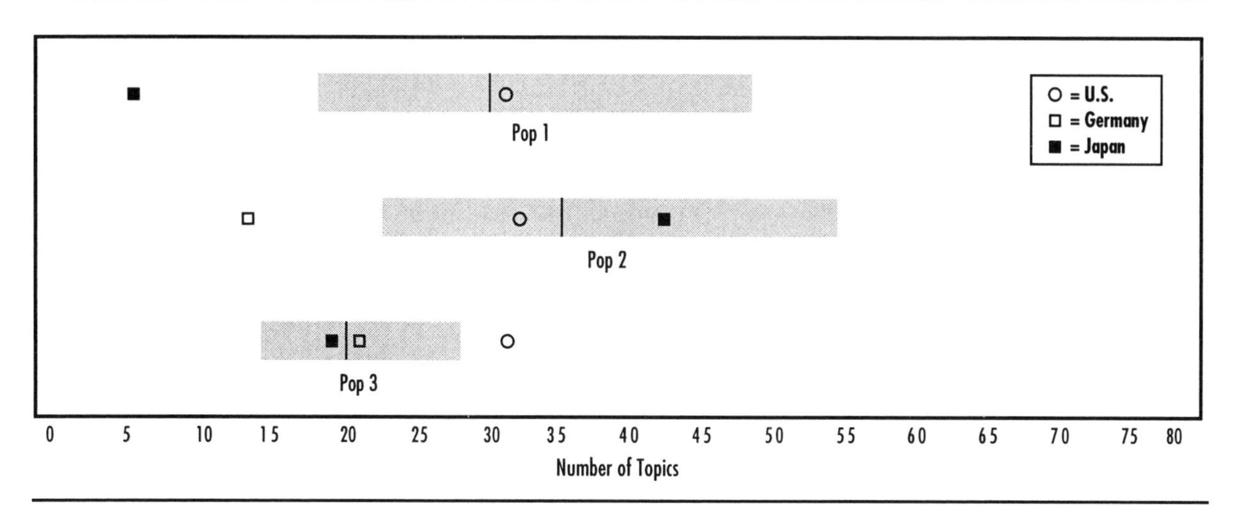

courses. However, the composite curricula are likely more representative of the aggregation to national composite curricula. Since testing will be of representative national samples, these aggregate or composite mathematics and science curricula may be the most relevant for explaining achievement differences.

Which Topics Do We Plan to Cover at the Student Populations TIMSS is Testing?

Now that we have focused on key grades in which we analyzed large samples, we can look more closely at specific topics. Suppose we constructed a composite "world" core curriculum based on those common topics intended by most countries in mathematics or science. How would the U.S. composite curriculum compare to this international composite curriculum?

Population 1 mathematics. Exhibit 11 presents data comparing the U.S. and international composite mathematics curricula. The topics listed came from topics in the TIMSS curriculum analysis frameworks. We see by simple counting that the U.S. composite core was more extensive than the international composite core. This makes more concrete what we saw for grade 4 in which the U.S. intended to cover far more than the international 75th percentile of topics (see Exhibit 1). The U.S. list includes all the topics from the international list and several additional topics. From this comparison, the U.S. composite mathematics curriculum seems much richer and more varied than the international composite for this grade. However, both the U.S. and international composites cover roughly comparable amounts of instructional time. Given this, the very diversity of the U.S. list is further evidence that the U.S. composite mathematics curriculum is unfocused. Since we plan to teach so many different topics, we are likely unable to devote much time to any one topic.

These data reveal some other important facts. The international list contains three topics that are present widely only in curriculum guides and not textbooks. U.S. textbooks and curriculum guides support all three of these topics. Internationally, textbooks include one topic not commonly supported by curriculum guides. That topic is 'data representation and analysis,' which deals with organizing data, presenting it in simple graphs, and calculating simple descriptive statistics such as averages. Both U.S. textbooks and state curriculum guides widely include this topic. Internationally, curriculum guides appear to lead — or, at least, not to be consistently matched with — the content of textbooks. This is not true in the U.S.

The U.S. also has several topics that are widely present only in textbooks and not curriculum guides. The state curriculum guides in the U.S. are clearly more inclusive than their international counterparts. U.S. textbooks include even more topics than the state curriculum guides. Both U.S. textbooks and state curriculum guides appear here to be following an inclusive strategy of providing something from many topics. U.S. textbooks follow this cautious strategy even more completely than state curriculum guides (a point discussed further later).

Exhibit 11.

Common U.S. and international topics for Population 1 mathematics.

The U.S. composite mathematics curriculum as represented in state curriculum guides includes far more topics at this grade level than does the international composite. The same is true for the U.S. textbook composite. U.S. textbooks include several topics not commonly included in state curriculum guides for the U.S. [This exhibit lists topics from the TIMSS' mathematics framework intended by at least 70 percent of the countries (international) or at least 70 percent of sampled state guides (U.S.) for the upper grade of Population 1 (U.S. grade 4). Bold-face labels are more general categories which subsume more specific (non-bold-face) topics. We arranged both lists (U.S. and international) in three categories - topics listed only in 70 percent of curriculum guides, topics listed in 70 percent of both curriculum guides and textbooks, and topics listed only in 70 percent of the textbooks. Asterisks mark topics receiving more extensive textbook attention.]

International	U.S.
CURRICULUM GUIDES (Not In Textbooks)	
Numbers	
Fractions and Decimals	
Decimal Fractions	
Geometry: Position, Visualization & Shape	
2-D Geometry: Coordinate Geometry	
Geometry: Symmetry, Congruence & Similarity	
Transformations	
CURRICULUM GUIDES (Included in Textbooks)	
Numbers	**Numbers**
Whole Numbers	Whole Numbers
Meaning*	Meaning*
Operations*	Operations*
Properties of Operations	
Fractions and Decimals	Fractions and Decimals
Common Fractions	Common Fractions
	Decimal Fractions
	Estimation & Number Sense Concepts
	Estimating Computations
Measurement	**Measurement**
Units*	Units*
Perimeter, Area & Volume	Perimeter, Area & Volume
	Estimation & Errors
Geometry: Position, Visualization & Shape	**Geometry: Position, Visualization & Shape**
2-D Geometry: Basics	2-D Geometry: Coordinate Geometry
2-D Geometry: Polygons & Circles	2-D Geometry: Basics
	2-D Geometry: Polygons & Circles
	3-D Geometry
	Geometry: Symmetry, Congruence & Similarity
	Transformations
	Congruence & Similarity
	Functions, Relations, & Equations
	Patterns, Relations & Functions
	Equations & Formulas
	Data Representation, Probability, & Statistics
	Data Representation & Analysis
	Uncertainty & Probability

Exhibit 11. (cont'd)

Common U.S. and international topics for Population 1 mathematics.

International	U.S.
EXCLUSIVELY IN TEXTBOOKS	
	Numbers Whole Numbers Properties of Operations Fractions and Decimals Relation of Common & Decimal Fractions Integer, Rational & Real Numbers Negative Numbers, Integers & Their Properties Other Numbers & Number Concepts Number Theory Counting Estimation & Number Sense Estimating Quantity & Size Rounding & Significant Figures **Proportionality** Proportionality Concepts
Data Representation, Probability, & Statistics Data Representation & Analysis	
	Validation & Structure Validation and Justification Structuring and Abstracting **Other Content**

If we wish to use more extensive coverage of topics in textbooks as evidence of focus, only three topics receive more extensive coverage at Population 1. These three topics — 'whole numbers: meaning' (including numeration), 'whole numbers: operations,' and 'measurement: units' — receive more attention both internationally and in the U.S. Thus, we have some evidence of an arithmetic- and measurement-based curriculum at this grade both in the U.S. and internationally. However, the inclusion of many other topics limits how much focus these emphases imply. In any case, we see evidence of varied but unfocused curricula in state curriculum guides — a lack of focus unchanged by U.S. textbooks. We typically cover markedly more topics and with less focus than the international benchmark. In short, our state curriculum guides at this targeted grade are clearly less focused than those of most other countries.

Population 2 mathematics. The comparative picture is much the same for the upper grade of Population 2 (as Exhibit 12 shows). There are two important additional observations for this population. First, some arithmetic and 'measurement' content —not common internationally — continues to persist in U.S. state curriculum guides and textbooks. U.S. textbooks and state curriculum guides continue to include 'whole numbers: operations,' 'common fractions,' 'percentages,' 'measurement: units' and 'measurement: estimation and errors' (the latter two not typically in U.S. algebra books). This is not true internationally (attention to these topics typically having been completed in earlier grades). Second, many mathematics topics, especially some prominent in U.S. mathematics reform proposals, receive attention that they do not receive internationally, both in algebra and "nonalgebra" textbooks and in state curriculum guides. These include topics in 'estimation and number sense,' 'proportionality concepts,' and topics in data representation and probability.

The U.S. composite curriculum and state mathematics curriculum guides at this level intend coverage of more topics than is typical internationally. These topics have two distinct sources — the persistence of arithmetic and 'measurement' topics introduced early but remaining (presumably because of ideas of distributed mastery), and new topics introduced at this level most likely because of reform concerns in U.S. mathematics education.

Both reasons seem in one sense justified. However, the result is curricula not focused on strategic topics and curricula that divide classroom mathematics time among more than the usual number of topics. We will likely see the results of this unfocused, "a little of everything" approach in the comparative mathematics achievements of our students. At the very least, we must question whether it is feasible to continue a distributed mastery approach while at the same time introducing new topics based on reform concerns. The former causes topics to persist several grades; the latter adds new topics while we seem unable to delete older material. Unless U.S. students prove surprisingly able when faced with this variety of contents, some priorities undoubtedly need setting. Some priorities were suggested in the NCTM *Standards*.[8] It appears that reform suggestions for adding topics have been more widely implemented than the accompanying suggestions for de-emphasizing other topics.

Exhibit 12.

Common U.S. and international topics for Population 2 mathematics.

The U.S. composite mathematics curriculum as represented in state curriculum guides includes more topics at this grade level than does the international composite. The same is true for U.S. textbooks. This is true both for algebra course textbooks and those for more general mathematics courses. These latter may include some but not an entire course of algebra (although here called "nonalgebra textbooks"). U.S. textbooks include several topics not commonly included in state curriculum guides for the U.S. [This exhibit lists topics from the TIMSS' mathematics framework intended by at least 70 percent of the countries (international) or at least 70 percent of sampled state guides (U.S.) for the upper grade of Population 2 (U.S. grade 8). Bold-face labels are more general categories that subsume more specific (non-bold-face) topics. Both lists (U.S. and international) are arranged in three categories. These three are topics listed only in 70 percent of curriculum guides, topics listed in 70 percent of both curriculum guides and textbooks, and topics listed only in 70 percent of the textbooks. Asterisks mark topics receiving more extensive textbook attention.]

International	U.S.-Nonalgebra Textbooks	U.S.-Algebra Textbooks
	CURRICULUM GUIDES (Not In Textbooks)	
Number	**Number**	**Number**
Integer, Rational & Real Numbers		
Real Numbers, Their Subsets & Properties		
	Estimation & Number Sense Concepts	**Estimation & Number Sense Concepts**
		Estimating Quantity & Size
		Rounding & Significant Figures
		Estimating Computations
	Exponents & Orders of Magnitude	Exponents & Orders of Magnitude
		Measurement
		Units
		Estimation & Errors
	Geometry: Position, Visualization & Shape	**Geometry: Position, Visualization & Shape**
		3-D Geometry
	Vectors	Vectors
Geometry: Symmetry, Congruence & Similarity		
Congruence & Similarity		
Constructions Using Straightedge & Compass		
Proportionality		
Proportionality Concepts		
	CURRICULUM GUIDES (Included in Textbooks)	
Number	**Number**	**Number**
	Whole Numbers	**Whole Numbers**
	Operations	Operations
	Fractions & Decimals	**Fractions & Decimals**
	Common Fractions	Common Fractions
	Relationship of Common & Decimal Fractions	Relationship of Common & Decimal Fractions
	Percentages	Percentages
Integer, Rational & Real Numbers	**Integer, Rational & Real Numbers**	**Integer, Rational & Real Numbers**
Integers & Their Properties	Integers & Their Properties	Integers & Their Properties
Rational Numbers & Their Properties		
	Estimation & Number Sense Concepts	
	Estimating Quantity & Size	
	Rounding & Significant Figures	
	Estimating Computations	
Other Numbers & Number Concepts		
Exponents, Roots & Radicals		
Measurement	**Measurement**	**Measurement**
	Units	
Perimeter, Area & Volume	Perimeter, Area & Volume	Perimeter, Area & Volume
	Estimation & Errors	
Geometry: Position, Visualization & Shape	**Geometry: Position, Visualization & Shape**	**Geometry: Position, Visualization & Shape**
2-D Geometry: Coordinate Geometry	2-D Geometry: Coordinate Geometry	2-D Geometry: Coordinate Geometry*
2-D Geometry: Basics	2-D Geometry: Basics	2-D Geometry: Basics
2-D Geometry: Polygons & Circles	2-D Geometry: Polygons & Circles	2-D Geometry: Polygons & Circles
3-D Geometry	3-D Geometry	

Exhibit 12. (cont'd)

Common U.S. and international topics for Population 2 mathematics.

International	U.S.-Nonalgebra Textbooks	U.S.-Algebra Textbooks
Geometry: Symmetry, Congruence & Similarity	**Geometry: Symmetry, Congruence & Similarity**	**Geometry: Symmetry, Congruence & Similarity**
Transformations	Transformations	Transformations
	Congruence & Similarity	Congruence & Similarity
Proportionality	**Proportionality**	**Proportionality**
	Proportionality Concepts	Proportionality Concepts
Proportionality Problems		
Functions, Relations, & Equations	**Functions, Relations, & Equations**	**Functions, Relations, & Equations**
Patterns, Relations & Functions	Patterns, Relations & Functions	Patterns, Relations & Functions
Equations & Formulas	Equations & Formulas*	Equations & Formulas*
	Data Representation, Probability, & Statistics	**Data Representation, Probability, & Statistics**
	Data Representation & Analysis	Data Representation & Analysis
	Uncertainty & Probability	Uncertainty & Probability
	EXCLUSIVELY IN TEXTBOOKS	
Number	**Number**	**Number**
Whole Numbers	**Whole Numbers**	**Whole Numbers**
	Meaning	Meaning
Properties of Operations	Properties of Operations	Properties of Operations
Fractions & Decimals	**Fractions & Decimals**	**Fractions & Decimals**
Common Fractions		
Decimal Fractions	Decimal Fractions	Decimal Fractions
Relationships of Common & Decimal Fractions		
Percentages		
	Integer, Rational & Real Numbers	**Integer, Rational & Real Numbers**
	Rational Numbers & Their Properties	Rational Numbers & Their Properties
	Real Numbers, Their Subsets & Properties	Real Numbers, Their Subsets & Properties*
	Other Numbers & Number Concepts	**Other Numbers & Number Concepts**
	Binary arithmetic and/or Other Number Bases	
	Exponents, Roots & Radicals	Exponents, Roots & Radicals*
		Complex Numbers & Their Properties
	Number Theory	Number Theory
	Counting	
Measurement		
Units		
	Geometry: Symmetry, Congruence & Similarity	
	Constructions Using Straightedge & Compass	
	Proportionality	**Proportionality**
	Proportionality Problems	Proportionality Problems
	Slope & Trigonometry	Slope & Trigonometry
	Validation & Structure	**Validation & Structure**
	Validation and Justification	Validation and Justification
	Structuring and Abstracting	Structuring and Abstracting
	Other Content	**Other Content**

We should note in passing that mathematics textbooks appear to follow a more inclusive strategy. This is true both for the U.S. and internationally. Whole number and fractions topics not in curriculum guides persist widely in both U.S. and other countries' textbooks. In the U.S., we include several other (more advanced) topics in both algebra and nonalgebra textbooks. This is true even when curriculum guides do not include these topics. U.S. textbooks seem unlikely to add focus to official state mathematics curricula that are largely unfocused.

Population 1 science. The international composite science curriculum for Population 1 has more topics than the U.S. composite science curriculum found in state curriculum guides (see Exhibit 13 but ignore, for the moment, those topics found only in textbooks). We saw earlier (Exhibit 10) that the composite science curriculum for this grade level showed plans for about 30 topics commonly among the state curriculum guides.

Why do we find such a difference in the number of topics between Exhibit 10 and Exhibit 13? Exhibit 10 averaged the numbers of topics intended in the sampled state guides to a characteristic number of topics for the U.S. composite science curriculum. Exhibit 13 shows the particular topics that are *common* among state science curriculum guides. Apparently most states plan science curricula with large numbers of topics but not many of the *same* topics. (Exhibit 13 shows only the science topics common to state plans for at least 70 percent of the states sampled).

Why would states differ so much in the topics they planned to include in their opportunities for science learning? Unlike school mathematics, U.S. school science appears not widely organized for cumulative learning and distributed mastery. If school science was organized that way, the list of commonly intended science topics would be far longer. Alternatively, school science may be organized around common processes in doing science. These processes could be applied with different contents. They would *need* to be used with a variety of contents to show that the processes were general and not specific to particular science topics. Topics could vary with local preference. This would allow a strong core of common science processes but a limited number of commonly intended science topics. We discuss this hypothesis later. However, since the *number* of science topics covered in most state plans is still large, these curricula are unfocused unless content is treated as very peripheral to process.

School science topics vary more among states by state preference — with or without an emphasis on science processes. Why were there so few topics in common (to 70 percent or more of the states)? If specific topics are to some extent incidental, states may choose topics based on their own preferences and the topics' perceived relevance to local situations. Geographical, ecological, and other aspects of local conditions may make particular topics of more interest to some states than others, especially in 'life sciences' and 'earth sciences.' Several of the specific common topics ('physical properties of matter,' 'electricity,' 'light,' etc.) are from physical science. This is a discipline with more restricted content and organized incrementally more like mathematics.

Exhibit 13.

Common U.S. and international topics for Population 1 science.

Looking only at the common topics in state curriculum guides (not those in textbooks) the U.S. composite science curriculum in state curriculum guides includes fewer topics at this grade level than the international composite. While state curriculum guides commonly plan for around 30 science topics, they are not the same science topics in each state so the list of specific common topics is smaller. Since each state still plans for a comparatively large number of topics, the state science curricula at this level still remain unfocused. [This exhibit lists topics from the TIMMS' science framework intended by at least 70 percent of the countries (international) or at least 70 percent of sampled state guides (U.S.) for the upper grade of Population 1 (U.S. grade 4). Bold-face labels are more general categories that subsume more specific (non-bold-face) topics. Both lists (U.S. and international) are arranged in three categories. These are topics listed only in 70 percent of curriculum guides, topics listed in 70 percent of both curriculum guides and textbooks, and topics listed only in 70 percent of the textbooks. Asterisks mark topics receiving more extensive textbook attention.]

International	U.S.
CURRICULUM GUIDES (Not In Textbooks)	
Earth Sciences	
Earth Features	
Bodies of Water	
Life Sciences	
Human Biology & Health	
Environmental and Resource Issues Related to Science	
Pollution	
Conservation of Material & Energy Resources	
World Population	
Food Production, Storage	
CURRICULUM GUIDES (Included in Textbooks)	
Earth Sciences	
Earth Processes	
Weather & Climate	
Earth in the Universe	
Earth In The Solar System	
Life Sciences	**Life Sciences**
Diversity, Organization, Structure of Living Things	Diversity, Organization, Structure of Living Things
Plants, Fungi*	Plants, Fungi*
Animals*	Animals*
Organs, Tissues	Organs, Tissues
Interactions of Living Things	Interactions of Living Things
	Biomes and Ecosystems
Interdependence of Life	Interdependence of Life
Physical Sciences	**Physical Sciences**
Matter	Matter
Physical Properties of Matter	Physical Properties of Matter
Energy and Physical Processes	Energy and Physical Processes
Energy Types, Sources, Conversions	Energy Types, Sources, Conversions
	Light
	Electricity
Environmental and Resource Issues	
Conservation of Land, Water, and Sea Resources	

Exhibit 13. (cont'd)

Common U.S. and international topics for Population 1 science.

International	U.S.
EXCLUSIVELY IN TEXTBOOKS	
	Earth Science
	Earth Features
	Landforms
	Bodies of Water
	Earth Processes
	Weather & Climate
	Physical Cycles
	Earth In The Universe
	Earth in the Solar System *
	Planets in the Solar System
	Life Sciences
	Diversity, Organization, Structure of Living Things
	Other Organisms
	Cells
	Life Processes and Systems Enabling Life Functions
	Energy Handling
	Sensing, Responding
	Life Spirals, Genetic Continuity, Diversity
	Life Cycles
	Reproduction
	Evolution, Speciation, Diversity
	Interaction of Living Things
	Habitats & Niches
	Animal Behavior
	Human Biology and Health
	Nutrition
	Physical Sciences
	Matter
	Classification of Matter
	Chemical Properties
	Structure of Matter
	Atoms, Molecules, Ions
	Energy and Physical Processes
	Sound and Vibration
	Forces and Motion
	Types of Forces
	Science, Technology, and Mathematics
	Interactions of Science, Mathematics, and Technology
	Applications of Science in Mathematics, Technology
	Interactions of Science, Technology, and Society
	Influence of Science, Technology on Society
	History of Science & Technology
	Environmental and Resources Issues Related to Science
	Conservation of Land, Water, and Sea Resources
	Conservation of Material & Energy Resources
	Food production, Storage
	Nature of Science
	Nature of Scientific Knowledge
	The Scientific Enterprise
	Science and Other Disciplines
	Science and Mathematics
	Science and Other Disciplines

Population 2 science. The findings for state science curricula in the upper grade of Population 2 (grade 8 in the U.S.) are similar to those for Population 1. (Compare Exhibit 14 for Population 2 science to Exhibit 13 for Population 1 science.) The number of topics in the composite science curriculum for this grade is far higher than the numbers for other countries. However, the *specific* topics common among state curriculum guides are fewer (compare Exhibit 14 to Exhibit 10). The explanations and possibilities are the same as those discussed earlier.

There is another, outstanding common feature to both grades (see Exhibit 13 and Exhibit 14). For both grades, science textbooks commonly included many topics not also in the corresponding state curriculum guides. This is in marked contrast to the international composite for science. U.S. science textbook content is clearly based on a more cautious, inclusive strategy (discussed further below). The lack of science topics widely in curriculum guides but not in textbooks supports this idea. Unlike science, we found mathematics topics in curriculum guides that were not in textbooks. Perhaps this is because the reform efforts in mathematics have had a longer time to make their impacts on mathematics curriculum guides and textbooks. Even when not also included in textbooks, some reform emphases were widely common in state mathematics guides. Soon the situation may be similar for science guides.

What Do We Expect Students to Do With the Content We Plan to Cover?

We also need to discuss the characteristic kinds and levels of mastery expected for students in the composite and state curriculum guides. We have considered only content so far. Typically, however, we include in curriculum plans both what content we intend students to encounter and also what we expect students to become able to do with that content. The TIMSS frameworks included both lists of specific content topics and lists of "performance" expectations. That is, they listed the kinds of tasks we might expect students to be able to perform for specific contents.

We developed performance expectations as expected, performable tasks rather than cognitive processes. Specific cognitive or thought processes might lie behind the performance of these tasks. However, characteristic thought processes are more closely tied to the culture in which education occurs. Specific tasks in school science and mathematics are more likely to be widely common among countries and less tied to specific cultures. Thus, they are more appropriate for cross-national comparisons (which are also necessarily multicultural). Broader categories organized related and similar specific performance expectations. Broader categories included 'investigating and problem solving,' 'mathematical reasoning,' 'investigating the natural world,' 'communicating,' etc.

When we analyzed curriculum guides and textbooks for the targeted grades, we examined both content and performance expectations. We identified one or more content topics for each

Exhibit 14.

Common U.S. and international topics for Population 2 science.

The U.S. composite science curriculum in state curriculum guides includes fewer topics than does the international composite at this grade level. We see this by looking only at the common topics in state curriculum guides and not those in textbooks. While state curriculum guides commonly plan for about 30 science topics, they are not the same science topics in each state. The list of specific common topics is thus smaller. The state science curricula are still unfocused since each state still plans for a comparatively large number of topics. [This exhibit lists topics from the TIMSS' science framework intended by at least 70 percent of the countries (international) or at least 70 percent of sampled state guides (U.S.) for the upper grade of Population 2 (U.S. grade 8). Bold-face labels are more general categories that subsume more specific (non-bold-face) topics. Both lists (U.S. and international) are arranged in three categories - topics listed only in 70 percent of curriculum guides, topics listed in 70 percent of both curriculum guides and textbooks, and topics listed only in 70 percent of the textbooks. Asterisks mark topics receiving more extensive textbook attention.]

International	U.S.
CURRICULUM GUIDES (Not In Textbooks)	
Life Sciences	
Diversity, Organization, Structure of Living Things	
Other Organisms (Types of Microorganisms)	
Cells	
Life Spirals, Genetic Continuity, Diversity	
Evolution, Speciation, Diversity	
Interactions of Living Things	
Biomes & Ecosystems	
Habitats & Niches	
Animal Behavior	
Human Biology & Health	
Physical Sciences	
Structure of Matter	
Subatomic Particles	
Energy and Physical Processes	
Magnetism	
Environmental and Resource Issues	
World Population	
Food Production, Storage	
Effects of Natural Disasters	
Nature of Science	
Nature of Scientific Knowledge	
CURRICULUM GUIDES (Included in Textbooks)	
Earth Sciences	
Earth Features	
Rocks, Soil	
Earth Processes	
Weather & Climate	
Life Sciences	**Life Sciences**
Diversity, Organization, Structure of Living Things	Diversity, Organization, Structure of Living Things
Plants, Fungi	Plants, Fungi
Animals	
Organs, Tissues*	Organs, Tissues
Life Processes and Systems Enabling Life Functions	
Energy Handling	
Sensing & Responding	

Exhibit 14. (cont'd)

Common U.S. and international topics for Population 2 science.

International	U.S.
Life Spirals, Genetic Continuity, Diversity	
Life Cycles	
Reproduction	
Interactions of Living Things	Interactions of Living Things
Interdependence of Life	Interdependence of Life
Human Biology and Health	
Disease	
Physical Sciences	**Physical Sciences**
Matter	Matter
Classification of Matter	
Physical Properties	Physical Properties
Chemical Properties	Chemical Properties
Structure of Matter	
Atoms, Ions, Molecules	
Energy and Physical Processes	Energy and Physical Processes
Energy Types, Sources, Conversions	Energy Types, Sources, Conversions
Heat & Temperature	
Light	
Electricity	Electricity
Chemical Transformations	
Explanations of Chemical Changes	
Forces and Motion	Forces and Motion
Types of Forces	
	Time, Space, and Motion
Science, Technology, and Mathematics	
Interactions of Science, Mathematics and Technology	
Science Applications in Math, Technology	
Interactions of Science, Technology and Society	
Influence of Science, Technology on Society	
Environmental and Resource Issues	**Environmental and Resource Issues**
Pollution	Pollution
Conservation of Land, Water, & Sea Resources	
Conservation of Material & Energy Resources	
	Nature of Science
	Nature of Scientific Knowledge
EXCLUSIVELY IN TEXTBOOKS	
Earth Sciences	**Earth Sciences**
Earth Features	Earth Features
	Landforms
	Bodies of Water
	Atmosphere
Rocks, Soil	Rocks, Soil
	Ice Forms
	Earth Processes
	Weather & Climate
	Physical Cycles
	Earth's History
	Earth in the Universe

Exhibit 14. (cont'd)
Common U.S. and international topics for Population 2 science.

	Earth in the Solar System
	Planets in the Solar System
	Life Sciences
	Diversity, Organization, Structure of Living Things
	Animals
	Other Organisms
	Cells
	Life Processes and Systems Enabling Life Functions
	Energy Handling
	Sensing & Responding
	Life Spirals, Genetic Continuity, Diversity
	Life Cycles
	Reproduction
	Evolution, Speciation, Diversity
	Interaction of Living Things
	Biomes & Ecosystems
	Human Biology and Health
	Nutrition
	Disease
Physical Sciences	**Physical Sciences**
	Matter
	Classification of Matter
	Structure of Matter
	Atoms, Ions, Molecules
	Subatomic Particles
	Energy and Physical Processes
	Heat & Temperature
	Sound & Vibration
Physical Transformations	Physical Transformations
Physical Changes	Physical Changes
	Chemical Transformations
	Chemical Changes
	Nuclear Chemistry
Forces and Motion	Forces and Motion
	Types of Forces
Time, Space, & Motion	
	Science, Technology, and Mathematics
	Nature or Conceptions of Technology
	Interactions of Science, Mathematics and Technology
	Applications of Science in Math, Technology
	Interactions of Science, Technology, and Society
	Influence of Science, Technology on Society
History of Science & Technology	**History of Science & Technology**
	Environmental and Resources Issues
	Conservation of Land, Water, & Sea Resources
	Conservation of Material & Energy Resource
	Food Production, Storage
	Effect of Natural Disasters
	Nature of Science
	The Scientific Enterprise
	Science and Other Disciplines
	Scientific and Mathematics
	Science and Other Disciplines

small piece of a document. We also identified one or more performance expectations for that document segment. We characterized each document segment by its profile of contents and expectations. Since we identified contents and expectations for those contents for each small document segment, our method linked the two types of data. We conducted many of the same kinds of analyses for performance expectations as for content topics. Data already presented showed U.S. mathematics and science curricula as unfocused. These curricula planned to cover many topics in each grade and did not focus on mastering a few, strategic topics. This was true for the composite curricula and most individual state curricula. Do we show a similar lack of focus in what we expect students to do with the content they encounter?

State plans and U.S. textbooks include a greater number of specific expectations than are common internationally (see Exhibit 15 for Population 1 mathematics). Many specific expectations that internationally are only in curriculum guides are both included in U.S. state curriculum guidelines *and* supported by mathematics textbooks. State mathematics curriculum guides commonly include some specific expectations from every major category in the TIMSS frameworks. U.S. textbooks include even more. U.S. textbooks provide for additional specific expectations not common to most state curriculum guides.

Considering both U.S. curriculum guides and textbooks, the U.S. includes every international commonly intended performance expectation. However, the picture changes when we restrict our attention only to the expectations emphasized (by more extensive textbook space). The few emphasized performance expectations are all from the categories of 'knowing' and 'using routine procedures.' This is true also internationally.

Are U.S. performance expectations "focused"? Probably not, since we again include more expectations than is common internationally. On the positive side, however, we pay at least some attention to the higher expectations common to the reform literature such as the 1989 NCTM *Standards*.[9] This is especially true if we consider both textbooks and curriculum guides. We may not reflect thoroughly yet the mathematics content recommendations of recent reform (or do so only by adding them to existing content and further decrease focus). However, we have included more thoroughly the higher performance expectations characteristic of those reform recommendations.

We devote little U.S. textbook space (less than 6 percent of the analyzed textbook segments) to all but the simplest performance expectations. We must question whether textbooks seriously incorporate the more diverse array of performances or merely mention them occasionally. The import of Exhibit 15 is a kind of "good news, bad news" story. We may attend to the more demanding performance expectations suggested by recent mathematics reform efforts, but we are unfocused. Alternatively, we focus but pay only "lip service" to more demanding expectations from recent reform recommendations.

Exhibit 15.

Common U.S. and international Population 1 mathematics performance expectations.

U.S. state curriculum guides and textbooks had a larger number of common performance expectations than their international counterparts. State curriculum guides included performance expectations from all major categories. U.S. textbooks included even more. [This exhibit lists topics from the TIMSS' mathematics framework intended by at least 70 percent of the countries (international) or at least 70 percent of sampled state guides (U.S.) for the upper grade of Population 1 (U.S. grade 4). Bold-face labels are more general categories that subsume more specific (non-bold-face) performance expectations. Both lists (U.S. and international) are arranged in three categories — specific expectations listed only in 70 percent of curriculum guides, those listed in 70 percent of both curriculum guides and textbooks, and those listed only in 70 percent of the textbooks. Asterisks mark performance expectations receiving more extensive textbook attention.]

International	U.S.
CURRICULUM GUIDES (Not in Textbooks)	
Investigating Problem Solving	
Predicting	
Verifying	
Mathematical Reasoning	
Developing notation & vocabulary	
Developing algorithems	
CURRICULUM GUIDES (Included in Textbooks)	
Knowing	**Knowing**
Representing	Representing
Recognizing equivalents	
Recalling mathematical objects and properties	Recalling mathematical objects and properties*
Using Routine Procedures	**Using Routine Procedures**
Using equipment	Using equipment
Performing routine procedures*	Performing routine procedures*
Using more complex procedures	Using more complex procedures*
Investigating & Problem Solving	**Investigating & Problem Solving**
Formulating and clarifying problems and solutions	Formulating and clarifying problems and solutions
Developing strategy	Developing strategy
Solving	Solving
	Predicting
	Verifying
	Mathematical Reasoning
	Generalizing
	Conjecturing
	Justifying and proving
	Communicating
	Using vocabulary and notation
	Relating representations
	Describing/discussing

Exhibit 15. (cont'd)
Common U.S. and international Population 1 mathematics performance expectations.

International	U.S.
EXCLUSIVELY IN TEXTBOOKS	
	Knowing Recognizing equivalents **Mathematical Reasoning** Developing notation and vocabulary Developing algorithms **Communicating** Critiquing

For Population 2 mathematics, the U.S. expectations are more like the common international expectations (see Exhibit 16). These expectations are more extensive for Population 2 mathematics than for Population 1. Again, textbooks emphasize only a few specific expectations, both internationally and in the U.S. However, they no longer focus only on 'knowing' and 'using routine procedures.' Emphasized expectations also include solving problems and some aspects of 'communicating.' The same dilemma occurs. The books either reflect reform demands for higher expectations but are unfocused or vice versa. In this case, however, the U.S. materials differ less from the international.

The number of specific expectations, for the U.S. and the international composite, are comparable for Population 1 science if we ignore expectations found only in textbooks. (See Exhibit 17, the science counterpart of Exhibit 15.) However, the only common expectations in U.S. state curriculum guides come from categories of 'understanding' and 'using tools, routine procedures, and science processes.' Internationally, science curriculum guides include 'conducting investigations' and 'interpreting investigational data' from the category 'investigating the natural world.' The sampled U.S. state science guides did not widely include these latter expectations. U.S. textbooks again included more varied and demanding performance expectations than did the state curriculum guides. 'Understanding: simple information' was the only common expectation emphasized, both in the U.S. and internationally.

The Population 1 state science curricula seem more focused, requiring fewer different expectations than their mathematics counterparts. (Look only at curriculum guides in Exhibit 17.) However, the state guides achieve this by lower expectations for what students are to do with science content. This does not differ significantly from what is common internationally. Of course, some state science curricula, just as the national curricula of some countries, may well include more demanding performance expectations. However, if so, these demanding performance expectations are not the same across states or across countries. Further, the number of possibilities for differing performance expectations is considerably more limited than the number of possibilities for differing content topics. Thus, if there are specific demands set out in some, but not 70 percent, of the state curricula, there are likely at most a few such expectations. As before, common science textbook material at this grade level is more inclusive showing a greater variety of differing common performance expectations, including some that are more demanding.

In Population 2 science, unlike Population 1 science, our states' science curriculum guides require fewer specific expectations than their international counterparts. (Compare Exhibit 18 to Exhibit 17.) If we include those found commonly only in U.S. textbooks, we see more, varied, and more demanding expectations. Considering only curriculum guides, only limited expectations are common among the states. They are, however, more demanding than was true for Population 1 science. Yet, the single emphasized common expectation is 'understanding simple information,' the least demanding performance expectation in the TIMSS science framework.

Exhibit 16.

Common U.S. and international Population 2 mathematics performance expectations.

U.S. and international curriculum guides and textbooks have comparable numbers of common performance expectations. State curriculum guides include performance expectations from all major categories. There are even more in textbooks. [This exhibit lists topics from the TIMSS' mathematics framework intended by at least 70 percent of the countries (international) or at least 70 percent of sampled state guides (U.S.) for the upper grade of Population 2 (U.S. grade 8). Bold-face labels are more general categories that subsume more specific (non-bold-face) performance expectations. Both lists (U.S. and international) are arranged in three categories - specific expectations listed only in 70 percent of curriculum guides, those listed in 70 percent of both curriculum guides and textbooks, and those listed only in 70 percent of the textbooks. Asterisks mark performance expectations receiving more extensive textbook attention.]

International	U.S. non-Algebra Textbooks	U.S. Algebra Textbook
	CURRICULUM GUIDES (Not in Textbooks)	
Mathematical Reasoning Developing notation and vocabulary		**Mathematical Reasoning** Developing notation and vocabulary
	CURRICULUM GUIDES (Included in Textbooks)	
Knowing Representing Recognizing equivalents Recalling mathematical objects and properties*	**Knowing** Representing	**Knowing** Representing
Using Routine Procedures Using equipment Performing routine procedures* Using more complex procedures	**Using Routine Procedures** Using equipment Performing routine procedures* Using more complex procedures*	**Using Routine Procedures** Using equipment Performing routine procedures* Using more complex procedures*
Investigating and Problem Solving Formulating and clarifying problems and situations Developing strategy Solving Predicting Verifying	**Investigating and Problem Solving** Formulating and clarifying problems and situations Developing strategy Solving* Predicting Verifying	**Investigating and Problem Solving** Formulating and clarifying problems and situations Developing strategy Solving* Predicting Verifying
Mathematical Reasoning Generalizing Conjecturing Justifying and proving	**Mathematical Reasoning** Developing notation and vocabulary Generalizing Conjecturing Justifying and proving	**Mathematical Reasoning** Generalizing Conjecturing Justifying and proving
Communicating Using vocabulary and notation Relating representations Describing/discussing	**Communicating** Using vocabulary and notation Relating representations Describing/discussing*	**Communicating** Using vocabulary and notation Relating representations Describing/discussing*
	EXCLUSIVELY IN TEXTBOOKS	
	Knowing Recognizing equivalents Recalling mathematical objects and properties*	**Knowing** Recognizing equivalents Recalling mathematical objects and properties*
Mathematical Reasoning Generalizing	**Mathematical Reasoning** Developing algorithms	**Mathematical Reasoning** Developing algorithms
Communicating Relating representations	**Communicating** Critiquing	**Communicating** Critiquing

Exhibit 17.

Common U.S. and international Population 1 science performance expectations.

U.S. and international curriculum guides and textbooks have comparable numbers of common performance expectations. State curriculum guides include performance expectations from only two major categories. U.S. textbooks include more. [This exhibit lists topics from the TIMSS' science framework intended by at least 70 percent of the countries (international) or at least 70 percent of sampled state guides (U.S.) for the upper grade of Population 1 (U.S. grade 4). Bold-face labels are more general categories that subsume more specific (non-bold-face) performance expectations. Both lists (U.S. and international) are arranged in three categories — specific expectations listed only in 70 percent of curriculum guides, those listed in 70 percent of both curriculum guides and textbooks, and those listed only in 70 percent of the textbooks. Asterisks mark performance expectations receiving more extensive textbook attention.]

International	U.S.
CURRICULUM GUIDES (Not in Textbooks)	
Investigating the natural world	**Using tools, routines procedures, and science processes**
Conducting investigations	Interpreting data
Interpreting investigational data	
CURRICULUM GUIDES (Included in Textbooks)	
Understanding	**Understanding**
Simple information*	Simple information*
Complex information	
Using tools, routine procedures, and science processes	**Using tools, routine procedures, and science processes**
Using apparatus, equipment, computers	Using apparatus, equipment, computers
Conducting routine experimental operations	
Gathering data	Gathering data
	Organizing and representing data
EXCLUSIVELY IN TEXTBOOKS	
Using tools, routine procedures, and science processes	**Understanding**
Organizing and representing data	Complex information
	Theorizing, analyzing, and solving problems
	Applying scientific principles to develop explanations
	Constructing, interpreting, and applying models
	Using tools, routine procedures, and science processes
	Conducting routine experimental operations
	Investigating the natural world
	Designing investigations
	Conducting investigations
	Interpreting investigational data
	Communicating
	Accessing and processing information
	Sharing information

Exhibit 18.

Common U.S. and international Population 2 science performance expectations.

For Population 2 science, the U.S. has fewer common performance expectations compared to those common internationally. State curriculum guides include performance expectations from several major categories. U.S. textbooks include even more. The only emphasized expectation is the least demanding. [This exhibit lists topics from the TIMSS' science framework intended by at least 70 percent of the countries (international) or at least 70 percent of sampled state guides (U.S.) for the upper grade of Population 2 (U.S. grade 8). Bold-face labels are more general categories that subsume specific (non-bold-face) performance expectations. Both lists (U.S. and international) are arranged in three categories — specific expectations listed only in 70 percent of curriculum guides, those listed in 70 percent of both curriculum guides and textbooks, and those listed only in 70 percent of the textbooks. Asterisks mark performance expectations receiving more extensive textbook attention.]

Intentions in TIMSS Countries	U.S. Intentions
CURRICULUM GUIDES (Not in Textbooks)	
Theorizing, analyzing, and solving problems	
Making decisions	
Investigating the natural world	**Investigating the natural world**
Identifying questions to investigate	Interpreting investigational data
Designing investigations	
Formulating conclusions from investigational data	
Communicating	
Sharing information	
CURRICULUM GUIDES (Included in Textbooks)	
Understanding	**Understanding**
Simple information*	Simple information*
Complex information*	
Thematic information	
Theorizing, analyzing, and solving problems	**Theorizing, analyzing, and solving problems**
Abstracting & deducing scientific principles	Constructing, interpreting, and applying models
Applying scientific principles to solve quantitative science problems	
Applying scientific principles to develop explanations	
Constructing, interpreting, and applying models	
Using tools, routine procedures, and science processes	**Using tools, routine procedures, and science processes**
Using apparatus, equipment, computers	Using apparatus, equipment, computers
Conducting routine experimental operations	Gathering data
Gathering data	Organizing and representing data
Organizing and representing data	
Interpreting data	Interpreting data
Investigating the natural world	
Conducting investigations	
Interpreting investigational data	
Communicating	
Accessing & processing information	
EXCLUSIVELY IN TEXTBOOKS	
	Understanding
	Complex information
	Theorizing, analyzing, and solving problems
	Applying scientific principles to solve quantitative problems
	Applying scientific principles to develop explanations
	Making decisions
	Investigating the natural world
	Designing investigations
	Conducting investigations
	Communicating
	Accessing and processing information

Do U.S. state science intentions and those of the U.S. composite science curriculum reflect a science process approach? That is, do we apply central processes to different contents? This would show the generality of the processes but lead to fewer widely common topics since particular topics would be largely incidental to process-oriented curriculum plans. We can likely decisively dismiss this possibility. Only a few performance expectations are common and those few are relatively undemanding. One could not likely offer serious and demanding process-based approaches to science education without making greater demands than our data show.

The alternative explanation for limited common science topics now seems more likely. U.S. science education is not likely guided either by ideas of distributed mastery or by a science process orientation. Specific topics are probably chosen because of local interests. At the time documents were collected, there had been no time for science education reform efforts to have impact on science curriculum *guides*. This was less true for mathematics reform efforts and mathematics curriculum guides, but the time was still short. In contrast, performance expectations in science *textbooks* seem to have foreshadowed some of the more demanding tasks that would be recommended in science reform efforts.

Curricula: Summary and Concluding Remarks

What main points emerge from this discussion of the U.S. composite curricula in science and mathematics and of the common topics and expectations from state curricula? Curricula in both areas are comparatively unfocused. They include far more topics than is common internationally. Topics, especially in mathematics, persist over several grades, much more so than internationally. This is true, but less so, for science. (Physical science topics are an exception and more closely resemble mathematics.) State curriculum guides agree considerably about mathematics topics but much less so for science topics. The range of performance expectations for mathematics is greater than for science. The expectations for mathematics — in curriculum guides, not textbooks — are more demanding than is commonly true for state science guides. The mathematics guides reflect expectations typical of mathematics education reform recommendations. Science guides do so only in a far more restricted way, but science reform documents had not been completed or widely disseminated at the time the documents sampled were gathered.

The net result, however, is that the U.S. composite curricula and most state curricula are unfocused in mathematics and in science, although less so. They deal with far more than the internationally typical numbers of topics. However, given roughly comparable amounts of instructional time, this topic diversity limits the average amount of time allocated to any one topic. Since it is unlikely that time is allocated evenly across topics, we can make a statement from these data only about average allocations and require more direct time data (e.g., from teachers) to compare the relative attention given to specific topics. These curricula express an intention to do something of everything and, on average, less of any one thing. In mathematics, this may be because of a model of distributed mastery over the grades. An implicit use of this model would cause topics to persist for more grades than the international norm. The reasons for the results in science are less clear but seem related to general science (content, not process) approaches that move from topic to topic.

The U.S. composite curricula respond somewhat to reform agendas in mathematics education. The major science reform documents were not extant at the time data were gathered for these investigations. Unfortunately, state curricula appear to add reform mathematics contents to traditional, persisting topics without a counter-balancing de-emphasis of topics. This is clearly contrary to the recommendations of the NCTM *Standards*. This inclusiveness helps result in even more unfocused curricula. Possibly this inclusiveness is a point along the way to a more thorough-going reform similar to NCTM recommendations and later stages will see the de-emphasis of topics as recommended and the consequent focusing of mathematics curricula plans.

The mathematics curricula do show a variety of performances, including many from reform agendas. However, they focus — if they focus at all — on more simple demands. This is even more true in science (which does not reflect on science reform efforts for obvious reasons). For science, there are far fewer common specific performance expectations, and the few included are among the least demanding.

The general impact of these unfocused curricula in mathematics and science likely includes lower "yields" from mathematics and science education in the U.S.; although, the exact results and the reasons for them will differ for mathematics and for science. While the relationship between curricular focus and achievement may remain conjectural, the conventional wisdom (e.g., about time on task) suggests that lower achievement results are likely and that the U.S. is likely to suffer in cross-national comparisons. Mathematics reform recommendations thus far seem to have affected curricula primarily by inclusion of additional topics, with a concomitant decrease in how focused the curricula are. This seems to result from our unwillingness to drop other topics when newer topics are added. The impact of science reform recommendations remains untested as yet. Reform agendas in mathematics have had some impact on how demanding the tasks are that students are expected to become able to perform.

Even were the evidence to suggest widespread use of more demanding tasks, demanding increasingly complex performances in curricula, which are unfocused — and trying to provide curricular experiences with too many content topics — seems destined not to produce the desired results. In short, to the extent that reform agendas in mathematics education have penetrated to widely reshape state curricula, they have done so thus far in a context of unfocused content coverage. This is likely to limit their success, in spite of the wisdom in these reform agendas and suggested curriculum standards.

A major fault seems to be our national, state, and local uncertainty about abandoning content topics. We seem willing to augment existing curricula with new topics, but less willing to abandon traditional topics. However, the new topics from reform agendas typically are coupled with more demanding, time-consuming, and complex performance requirements for students. These reform agendas put a premium on focused curricula — on paying careful attention to a smaller number of strategic topics — and include explicit recommendations about topics that should receive less emphasis. We must seriously question the practice of augmenting curricula with additional topics without paying the price of eliminating others. For mathematics, we must also seriously question the traditional approach of distributed mastery that has led to topics persisting over unusually many grades, further crowding and unfocusing our mathematics curricula. Unless strong empirical evidence supports content diversity as more productive than focused curricular attention for U.S. students, U.S. policy at all levels would seem to need reexamination of the effects and bases for maintaining traditional topics in an inclusive context at the price of curricular focus.

Chapter 2

U.S. TEXTBOOKS: CONFLICTING DEMANDS, CAUTIOUS VISIONS

Textbooks, at least as we typically use them in the U.S., serve as important bridges and tools. They are bridges between the worlds of plans and intentions, and of classroom activities shaped in part by those plans and intentions. Official plans, goals, visions, and intentions reside in official documents — documents that help guide and shape what happens in state and U.S. classrooms and that help describe curricular opportunities and how to distribute them among programs and grades. The actual creation of opportunities occurs in classrooms by teachers in their interactions with students.

Those two worlds — that of official intentions and that of actual classroom activities — are tied together, in part, by textbooks. Official intentions affect the content, organization, and design of textbooks. Topics, activities, and tasks present in textbooks as resources help shape what actually happens in classrooms. While teachers may select from or supplement textbooks, for many U.S. mathematics and science teachers, textbooks at a "micro" level *are* the curriculum that guides mathematics and science instruction.[10]

Textbooks make content available, organize it, supply it with suggested activities, provide supporting materials, and set out supporting tasks for students in one handy location and in a form designed to be appealing to students. They do this with sufficient detail to support day-to-day classroom instruction.[11] Teachers often plan lessons beginning with a segment of the selected textbook on which they reflect and build classroom activity plans cross-referenced to those textbooks.

In these very common cases, textbooks are, by default and overwhelming demand, the backbone of "micro" organization for the classroom activities. They provide the fine details of curriculum expressed more broadly and less directly supportive in official curriculum documents. Textbooks define the domain of *implementable* day-to-day curricular possibilities. Without restricting what teachers *may* choose to do, they drastically affect what U.S. teachers are *likely* to do under the pressure of daily instruction.

Textbooks have an important instrumental role in shaping the details of how teachers put into practice curricular goals and visions in their classrooms. Given this, we must ask, "Do U.S. mathematics and science textbooks add guidance and focus to help teachers cope with unfocused curricula?" Unfortunately, but for very understandable reasons, the answer is "no."

Just as we spoke of composite curricula, we will speak of the average and characteristic features of U.S. textbooks. We will examine composites of sampled textbooks, those used in instruction for the majority of U.S. students studying science and mathematics at various grade levels during 1992-93. We have no national, official textbooks. A competitive, free market shapes mathematics and science textbooks, as it does other U.S. textbooks. They are influenced by the real experiences of teachers, the advice of experts, and the content of official curriculum documents. At times and given time, they are also influenced by the agendas of reform movements in U.S. mathematics and science education, especially as these are reflected indirectly in state and local demands and in market forces.

We analyzed textbooks for the three student populations being tested by TIMSS. These books were not sampled randomly but by using the best available data on how widely each was used. We supplemented this basic sample with special samples to include important specialized textbooks — Algebra I texts commonly used at grade 8, single-discipline texts such as grade 8 earth science books, etc. We restricted Population 3 textbooks to those used by specialist populations — in particular, calculus and physics books.

How Many Topics Do U.S. Textbooks Typically Include?

What can we say about the characteristic features of these textbooks in the aggregate and of their relationship to unfocused official curricula seen in U.S. composites and across individual states? First, as with curriculum plans, U.S. textbooks include many more topics than do other countries' textbooks. This is true both in mathematics and, especially, science.

U.S. textbooks at all three populations in our sample contain remarkably large numbers of mathematics topics (see Exhibit 19). U.S. mathematics textbooks average 30 to 35 topics in Populations 1 and 2. Population 3 mathematics textbooks, used by advanced students in specialized courses, still include an average of 25 topics. This is far above the 75th percentile of TIMSS countries. Japan has 20 topics at Population 1 but fewer than 10 at both other populations. This is so quantitatively different that it certainly reflects qualitative differences as well. Germany, for which textbook data is available only for Population 2, is similar to Japan.

Is the situation better for U.S. science textbooks? Quite the opposite. At all three population levels, U.S. science textbooks included far more topics than even the 75th percentile internationally (see Exhibit 20). The differences are far greater than those for mathematics textbooks. They are far greater than Germany's or Japan's difference from the internationally typical. U.S. science textbooks include just over 50 to 65 topics. Japanese science textbooks offer less than 20, as does Germany (only Population 2 data available).

Exhibit 19.

Number of mathematics textbook topics.

U.S. textbooks included far more mathematics topics than typical internationally in all three grades for which we analyzed textbooks. This was in strong contrast to Germany and, especially, Japan. [The gray bars show how many topics, averaged across a country's textbooks, were in the mathematics books at the upper grades of Populations 1 and 2 (U.S. grades 4 and 8) and for the mathematics specialists of Population 3. The bars extend from the 25th percentile to the 75th percentile for the number of topics among countries. The black line indicates the median number of topics for each population. We marked the U.S., Germany, and Japan individually. German textbook data were not available for Populations 1 and 3.]

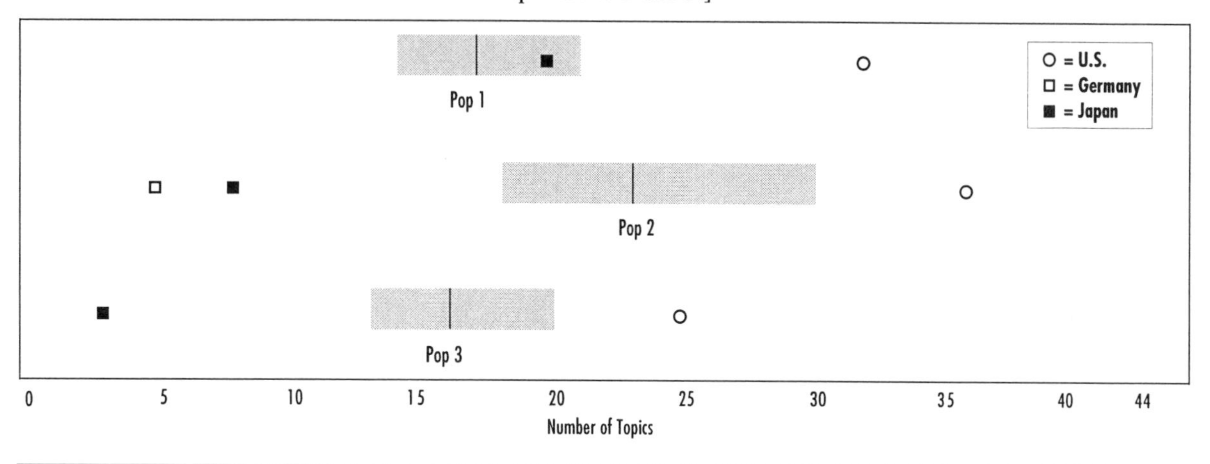

Exhibit 20.

Number of science textbook topics.

U.S. textbooks included far more science topics than typical internationally in all three grades for which we analyzed textbooks. The science differences are even larger than the mathematics differences. It was also in strong contrast to Germany and, especially, Japan. [The gray bars show how many topics, averaged across a country's textbooks, were in the science books at the upper grades of Populations 1 and 2 (U.S. grades 4 and 8) and for the science specialists of Population 3. The bars extend from the 25th percentile to the 75th percentile among countries in the number of topics for each population. The black line indicates the median number of topics for each population. We marked the U.S., Germany, and Japan individually. German textbook data were not available for Populations 1 and 3.]

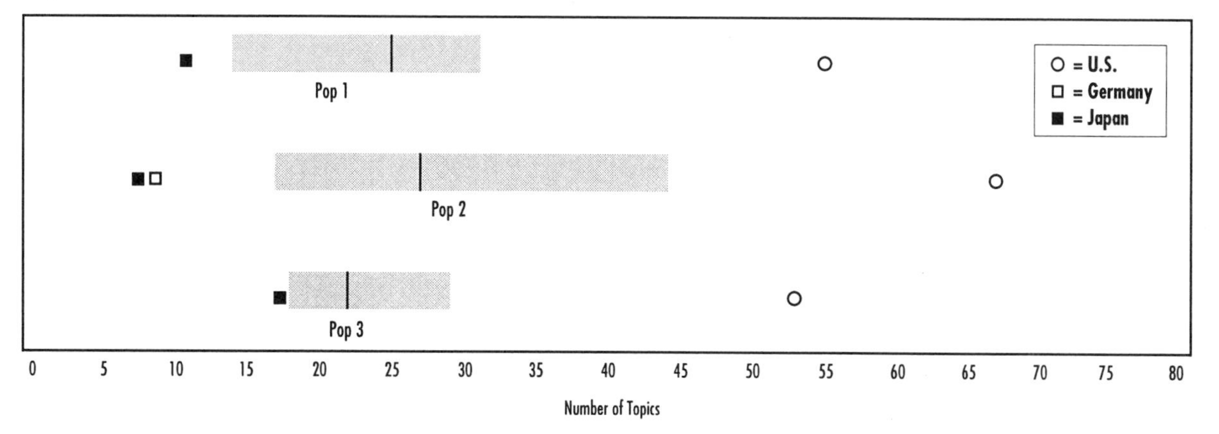

In both mathematics and science, U.S. textbooks included far more topics than the corresponding textbooks of other countries. This is true in both areas — science and mathematics — and for all three grade levels investigated. It is essentially impossible for textbooks so inclusive to help compensate for unfocused official curricula. U.S. textbooks support and extend the lack of focus seen in those official curricula.

How Much Space Do Textbooks Devote to the Few Topics Given the Most Attention?

We have seen that U.S. mathematics and science curriculum guides are unfocused. Can textbooks help this? How could textbooks including as many topics as the U.S. mathematics and science textbooks contribute focus to classroom instruction? To do so, they would have to devote disproportionately more space to a few key topics with far briefer "mentions" to the others. We need to know if this is true for U.S. mathematics and science textbooks.

We investigated this question by looking at the percentage of textbook space devoted to the few topics receiving the most space and attention. U.S. and other nations' teams partitioned textbooks into "blocks" (small segments used for analysis). The analysis focused on the percentage of these blocks. We chose to investigate the percentage of mathematics and science textbooks (blocks) devoted to the five topics receiving the most space and attention.

The U.S. grade 4 mathematics books on average are comparatively more focused than their numbers of topics suggest (see Exhibit 21). The five most emphasized topics account for, on average, about 60 percent of the U.S. mathematics books at this level. Three topics alone account for half of the average book — 'whole numbers: operations,' 'measurement: units' and 'common fractions.' The results are similar for Japan, although only two of the five topics are the same as the U.S.

Certainly we must consider the U.S. and Japanese textbooks arithmetic and algebra "driven." Internationally, typically the first five topics account on average for an even higher content percentage — over 85 percent. U.S. grade 4 books are inclusive (have many topics) but somewhat focused (space dominated by a few emphasized topics). Even here, however, they are not as much so as corresponding textbooks in other countries.

In contrast to mathematics textbooks at the same level, U.S. grade 4 science textbooks are as unfocused as the number of topics suggested. They are also more unfocused than their international counterparts (see Exhibit 22). The five most emphasized topics account for just over one-fourth of U.S. grade 4 science textbooks. This compares to almost three-fourths internationally and in Japan (and about 60 percent in grade 4 mathematics books). The *single* most emphasized science topic in the average of all TIMSS countries and in Japan accounts for almost as much of the typical science textbook as *all five* of the U.S.'s most emphasized topics. The inclusive nature of U.S. science textbooks suggested by examining the number of topics covered is not off-set by a less obvious focus on a few key topics, as it was in mathematics. These books truly do a little of almost everything and not much of any one thing.

Exhibit 21.

The five topics emphasized most in Population 1 mathematics textbooks.

At Population 1, U.S. mathematics books were somewhat focused but not as much so as was true internationally. For the upper grade of Population 1, the five most emphasized topics accounted for about 60 percent of U.S. and Japanese mathematics textbooks on average. This was far lower than the international average of over 85 percent. In the U.S., one of the emphasized "topics" was 'other content,' a combination of an unknown number of topics not in the TIMSS mathematics framework. [Data collection partitioned each textbook into "blocks" — small segments used as units of analysis. Each bar shows the cumulative percentage of textbook "blocks" for the five topics with the largest percentages of blocks devoted to them. These data are country or multi-country averages for mathematics textbooks at the upper grade of Population 1 (U.S. grade 4). Bars are for the average over all TIMSS countries, for Japan, and for the U.S. German textbook data were not available at this level. For the U.S. and Japan, each segment's specific TIMSS framework topic (e.g., 'whole number operations') could be determined and is shown. Internationally, the five topics varied among countries. We show them only as most emphasized (highest percentage of blocks), second most emphasized, etc.]

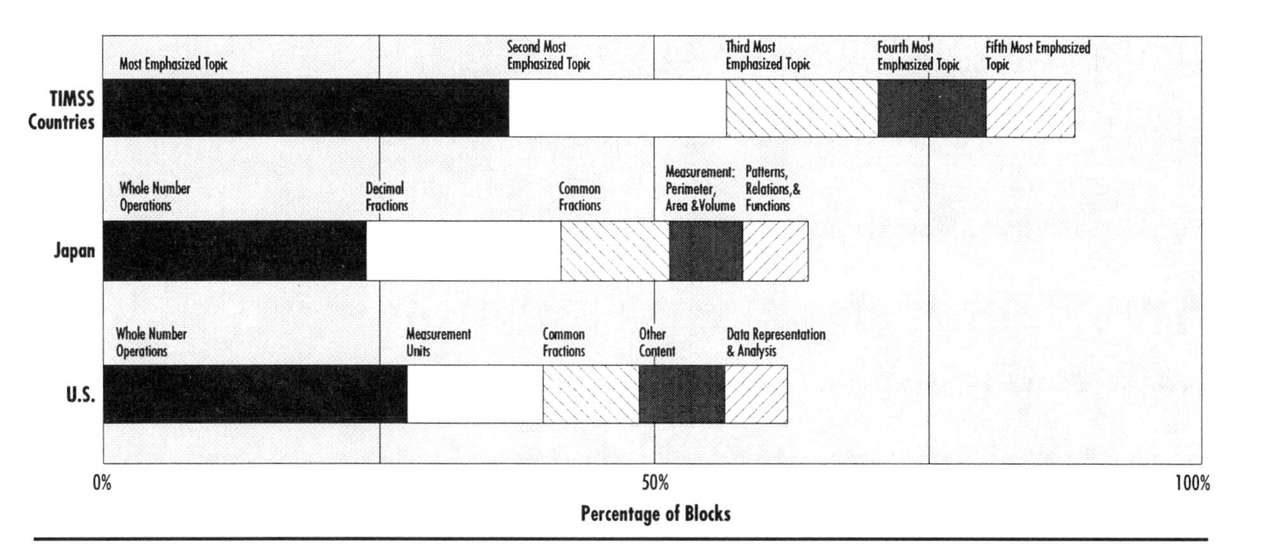

Exhibit 22.

The five topics emphasized most in Population 1 science textbooks.

At Population 1, U.S. science books were comparatively unfocused. On average the five most emphasized topics accounted for just over 25 percent of U.S. science textbooks compared to 70 to 75 percent internationally and for Japan. [Data collection partitioned each textbook into "blocks" — small segments used as units of analysis. Each bar shows the cumulative percentage of textbook "blocks" for the five topics with the largest percentages of blocks devoted to them. These data are country or multi-country averages for science textbooks at the upper grade of Population 1 (U.S. grade 4). Bars are for the average over all TIMSS countries, for Japan, and for the U.S. German textbook data were not available at this level. For the U.S. and Japan, each segment's specific TIMSS framework topic (e.g., 'whole number operations') could be determined and is shown. Internationally, the five topics varied among countries. We show them only as most emphasized (highest percentage of blocks), second most emphasized, etc.]

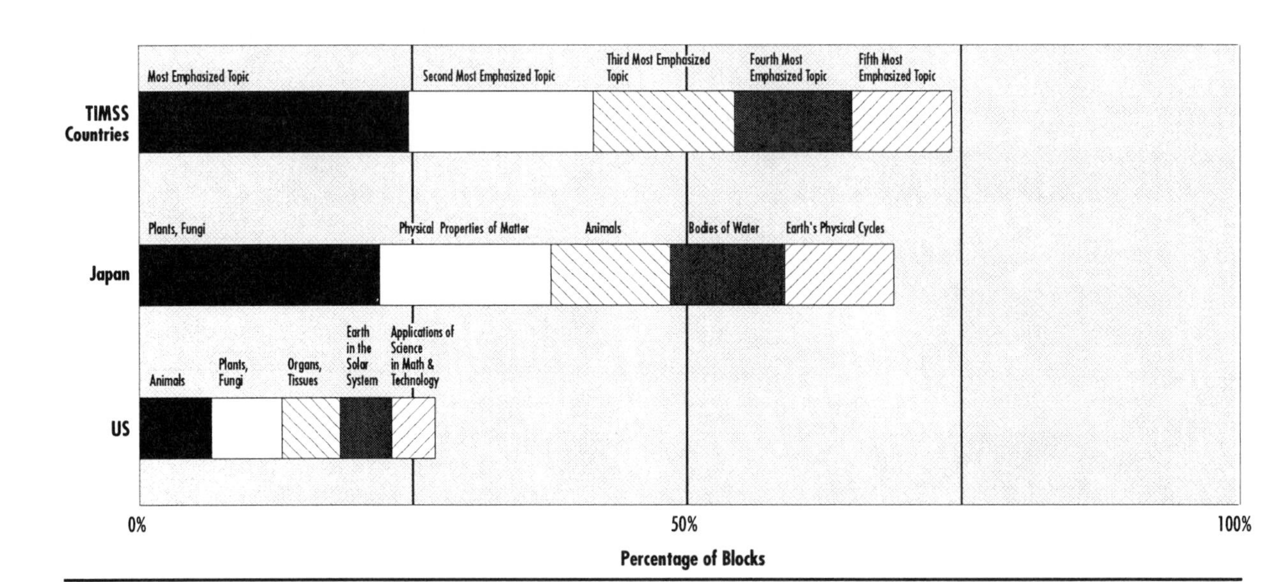

The corresponding data for grade 8 mathematics textbooks are very informative (see Exhibit 23).We analyzed books intended as texts for an Algebra I course separately from those intended for more general grade 8 mathematics courses. These general texts might include some algebra but were not for algebra courses. Here we call them "nonalgebra" texts to distinguish them from algebra course texts, even though they may include some algebra content.

Internationally, the top five topics account for about 75 percent of textbook space on average. This is true in Germany as well, and in Japan the top five topics account for over 80 percent of the space. By contrast, the five most emphasized topics account for less than 50 percent of the books' space in U.S. grade 8 books for more general courses. Even so, one of these topics is the catch-all 'other content' category that likely represents several other topics. These U.S. texts are clearly far less focused than their international Population 2 counterparts or than grade 4 U.S. mathematics books. In contrast, the five most emphasized topics completely dominate U.S. Algebra I books. These books are highly focused and very exclusive about content.

We actually have the top five topics summing to over 100 percent, which must appear mysterious at first. However, data collectors could label each "block" with as many topics as necessary to characterize its content. We did this to allow for more complex, integrated segments of textbooks. Analysts rarely assigned more than two categories and assigned two categories in only a small percentage of the cases. In the case of Algebra I books — which often combined equation content with other content (geometrical facts, proportionality, etc.) — "blocks" with two codes were enough to produce a total of more than 100 percent. This was the exception rather than the rule.

The specific emphasized topics in the U.S., Germany, and Japan are interesting. In both Japan and Germany algebra and geometry topics accounted for over 75 percent of the content. In U.S. Algebra I books, equations were about 75 percent of the content. Functions and exponent content further supplemented equations as algebra content. However, the nonalgebra U.S. textbooks contained only a mixture of contents, including small amounts of algebra, persisting arithmetic topics, and no geometry at all among the five most emphasized topics. Certainly the place of geometry was far different in the three countries, as was how focused the textbooks were.

U.S. grade 8 general or composite science textbooks (see Exhibit 24) follow the patterns similar to those for grade 8 mathematics books and grade 4 science books. The five most emphasized topics in grade 8 "general science" books, account for just over 50 percent of the content compared to an international average of about 60 percent. In Japan, almost 100 percent of the content is accounted for by the top five topics. Fortunately, about 70 percent of U.S. grade 8 students are instructed using single area textbooks — a physical science (including both physics and chemistry), life science, or earth science book — and these are far more focused than general science texts. In these textbooks, the top five topics accounted for almost

Exhibit 23.

The five topics emphasized most in Population 2 mathematics textbooks.

At Population 2, U.S. mathematics books, other than Algebra I textbooks, were much less focused than their international counterparts, or even than Population 1 mathematics textbooks. Only about half their content was accounted for by the five most emphasized topics in comparison to an international average of about 75 percent. One of the emphasized U.S. "topics" was again the catch-all 'other content' category that makes the degree of focus portrayed misleadingly high. U.S. Algebra I textbooks were highly focused with all content accounted for by the five most emphasized topics. [Data collection partitioned each textbook into "blocks" — small segments used as units of analysis. Each bar shows the cumulative percentage of textbook "blocks" for the five topics with the largest percentages of blocks devoted to them. These data are country or multi-country averages for mathematics textbooks at the upper grade of Population 2 (U.S. grade 8). Bars are for the average over all TIMSS countries, for Japan, Germany, and for the U.S. For the U.S., Germany, and Japan, each segment's specific TIMSS framework topic (e.g., 'equations and formulas') could be determined and is shown. Internationally, the five topics varied among countries. We show them only as most emphasized (highest percentage of blocks), second most emphasized, etc.]

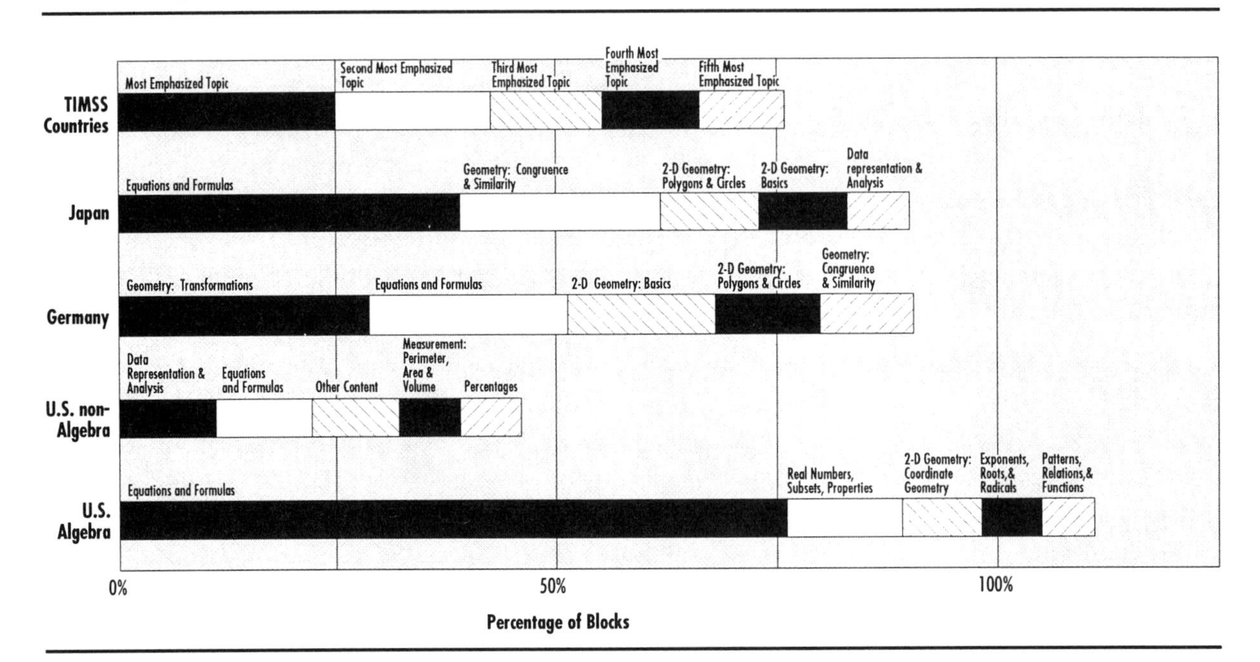

Exhibit 24.

The five topics emphasized most in Population 2 science textbooks.

At Population 2, U.S. science books, other than single-area textbooks, were much less focused than their international counterparts. About 50 percent of their content was accounted for by the five most emphasized topics compared to an international average of about 60 percent. US single-area textbooks in physical science, life science, or earth science were highly focused with the five most emphasized topics accounting for more of the books together than was true internationally. [Data collection partitioned each textbook into "blocks" — small segments used as units of analysis. Each bar shows the cumulative percentage of textbook "blocks" for the five topics with the largest percentages of blocks devoted to them. These data are country or multi-country averages for science textbooks at the upper grade of Population 2 (U.S. grade 8). Bars are for the average over all TIMSS countries, for Japan, Germany, and for the U.S. For the U.S., Germany, and Japan, each segment's specific TIMSS framework topic (e.g., 'rocks and soil') could be determined and is shown. Internationally, the five topics varied among countries. We show them only as most emphasized (highest percentage of blocks), second most emphasized, etc.]

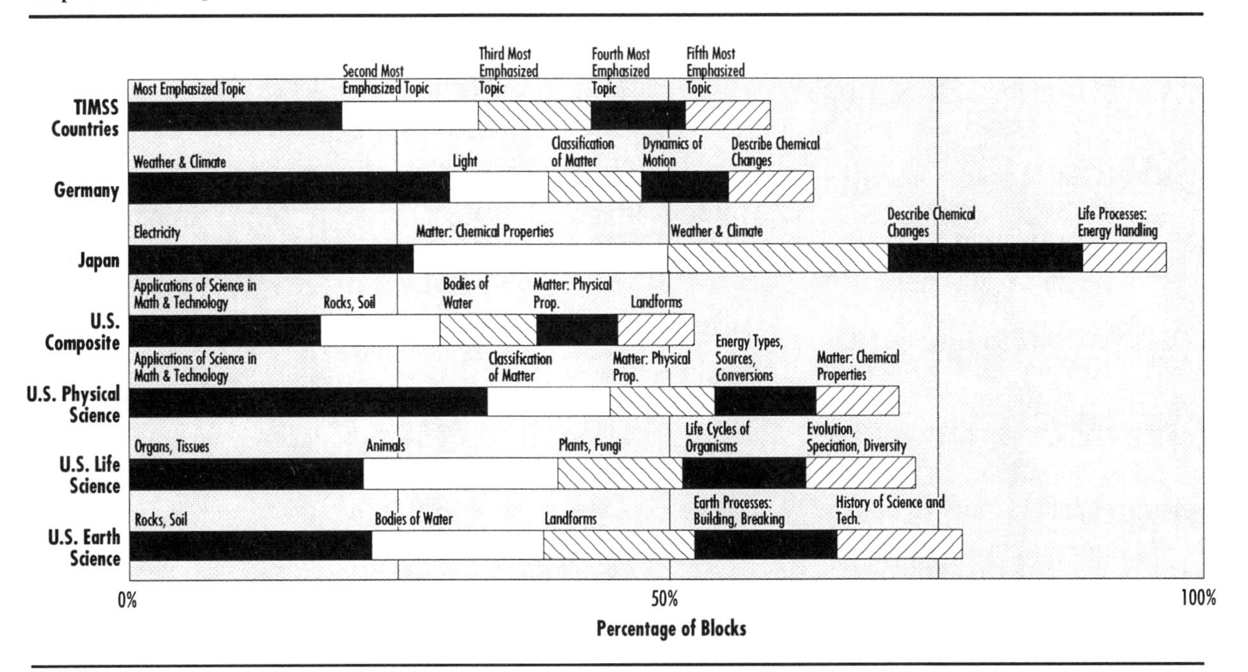

75 percent of the books' contents in all cases. In this way the single area science textbooks are like the grade 4 mathematics textbooks — inclusive (have many topics) but somewhat focused (space dominated by a few emphasized topics). However, since students using these textbooks receive instruction primarily in one area of science, their use raises concerns for aggregate achievement.

How and How Well Do U.S. Mathematics and Science Textbooks Reflect Reform Concerns?

We noted earlier that U.S. textbooks generally included more topics than did the corresponding curriculum guides (Exhibit 11 to Exhibit 14). In both science and especially mathematics, many of these textbook topics reflected the recommendations of reform agendas. Often, however, we simply add these additional topics to the existing array of topics while allowing more traditional content to persist. Thus, textbook publishers directed some attention to reform concerns but in an inclusive, unfocused way. (The data on textbooks' lack of focus presented earlier now supports this claim.)

Similarly, we saw that mathematics textbooks include more varied and demanding tasks for students (Exhibit 15 to Exhibit 18). However, these texts emphasize only more routine matters. Science textbooks included even less of the more demanding tasks suggested by reform efforts in science education. Both U.S. grade 8 science and mathematics textbooks seem to emphasize less demanding student performances comparatively more than the more demanding performance expectations (see Exhibit 25 and Exhibit 26). However, these textbooks compare favorably with their international counterparts. The mathematics textbooks seem to emphasize a greater variety of performances, including categories reflecting the concerns of mathematics reform efforts. Science textbooks do less so, perhaps because high profile U.S. mathematics reform reports — such as the NCTM *Standards* — have been in circulation longer than the corresponding science reports.

Textbooks: Summary and Concluding Remarks

U.S. textbooks in mathematics and, especially, in science include material on far more topics than is typical in other countries. U.S. science and mathematics textbooks tend to be "a mile wide and an inch deep." That is, although U.S. textbooks include many more topics than other countries' books, the few most emphasized topics account for less content than is the case internationally. Algebra I textbooks, single-area grade 8 science textbooks, and grade 4 mathematics textbooks are exceptions to this pattern. U.S. textbooks are inclusive — adding some of almost all contents but dropping little — and, like official curricula, they are largely unfocused. U.S. mathematics textbooks include a variety of performance expectations, including some of the more demanding ones. However, they more commonly emphasize the less demanding. U.S. science textbooks include less variety in performance expectations and emphasize even more heavily the less demanding.

Exhibit 25.

Textbook space devoted to major Population 2 science performance expectations.

U.S. textbook performance expectations in grade 8 science emphasized 'understanding' almost to the exclusion of all other expectations except 'using tools, routine procedures, and science processes.' This emphasis on understanding was typical internationally. [The bars represent the percentage of blocks devoted to each of the five most inclusive performance expectation categories in science textbooks for the upper grade of Population 2 (grade 8). Bars represent the average over all TIMSS countries and the U.S., Germany, and Japan specifically.]

Country	Understanding	Theorizing, Analyzing, and Solving Problems	Using Tools, Routine Procedures, & Science Processes	Investigating the Natural World	Communicating
Argentina					
Australia					
Austria					
Belgium (Fl) *					
Belgium (Fr) *					
Bulgaria					
Canada					
China, People's Rep. of					
Colombia					
Cyprus					
Czech Republic					
Denmark					
Dominican Republic					
France					
Germany					
Greece					
Hong Kong					
Hungary					
Iceland					
Iran					
Ireland					
Israel					
Italy					
Japan					
Korea					
Latvia					
Lithuania					
Mexico					
Netherlands					
New Zealand					

* The national Research Coordinators of Belgium have collected data only from curriculum guides. Due to the great level of detail of the guides, and their extensive use, data from these are compared in this display with the textbook data supplied from all other countries.

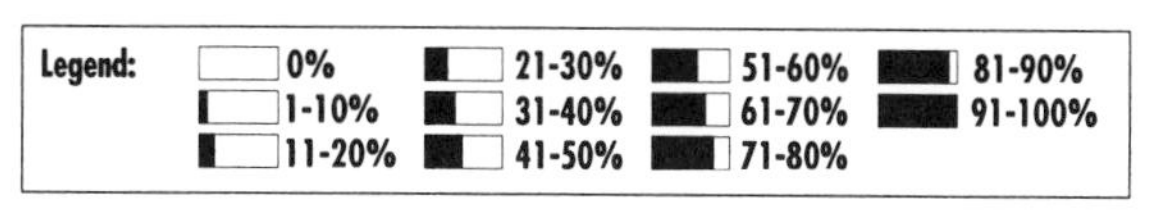

Note: Percentages across all categories do not sum to 100% because of the multiple codings associated with some blocks.

Missing data: Philippines, Thailand; Germany did not code Biology text.

Exhibit 25. (cont'd)

Textbook space devoted to major Population 2 science performance expectations.

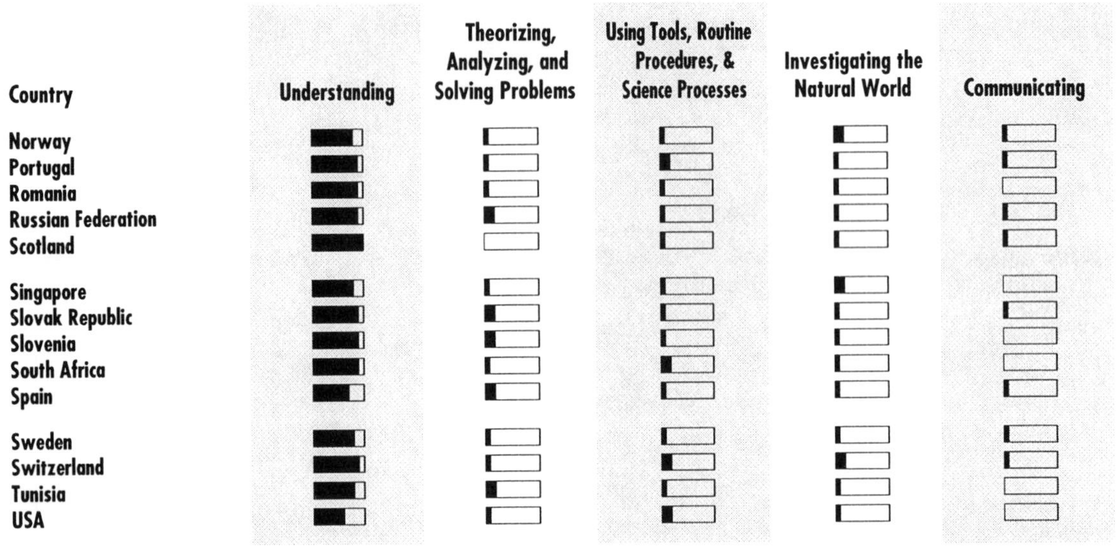

Country	Understanding	Theorizing, Analyzing, and Solving Problems	Using Tools, Routine Procedures, & Science Processes	Investigating the Natural World	Communicating
Norway					
Portugal					
Romania					
Russian Federation					
Scotland					
Singapore					
Slovak Republic					
Slovenia					
South Africa					
Spain					
Sweden					
Switzerland					
Tunisia					
USA					

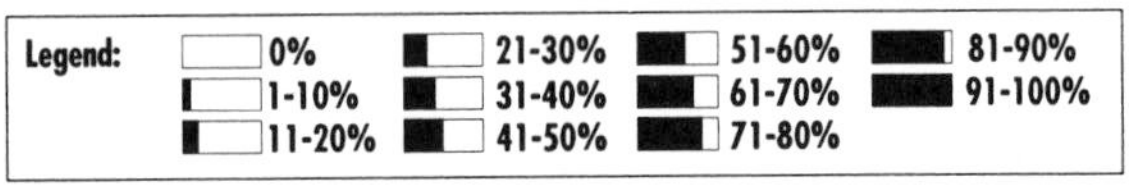

* The national Research Coordinators of Belgium have collected data only from curriculum guides. Due to the great level of detail of the guides, and their extensive use, data from these are compared in this display with the textbook data supplied from all other countries.

Note: Percentages across all categories do not sum to 100% because of the multiple codings associated with some blocks.

Missing data: Philippines, Thailand; Germany did not code Biology text.

Legend: 0%, 1-10%, 11-20%, 21-30%, 31-40%, 41-50%, 51-60%, 61-70%, 71-80%, 81-90%, 91-100%

Exhibit 26.

Textbook space devoted to major Population 2 mathematics performance expectations.

U.S. textbook performance expectations in grade 8 mathematics emphasized 'knowing' and 'using routine procedures' (including more complex procedures such as estimating and graphing) with some additional emphasis on communication. This pattern was more balanced than its international counterpart and may reflect more depth of attention to reform concerns to make more and more complex demands of students. Germany showed even more emphasis on 'knowing' but also more on 'investigating and problem solving.' [The bars represent the percentage of blocks devoted to each of the five most inclusive performance expectation categories in mathematics textbooks for the upper grade of Population 2 (grade 8). Bars represent the average over all TIMSS countries and the U.S., Germany, and Japan specifically.]

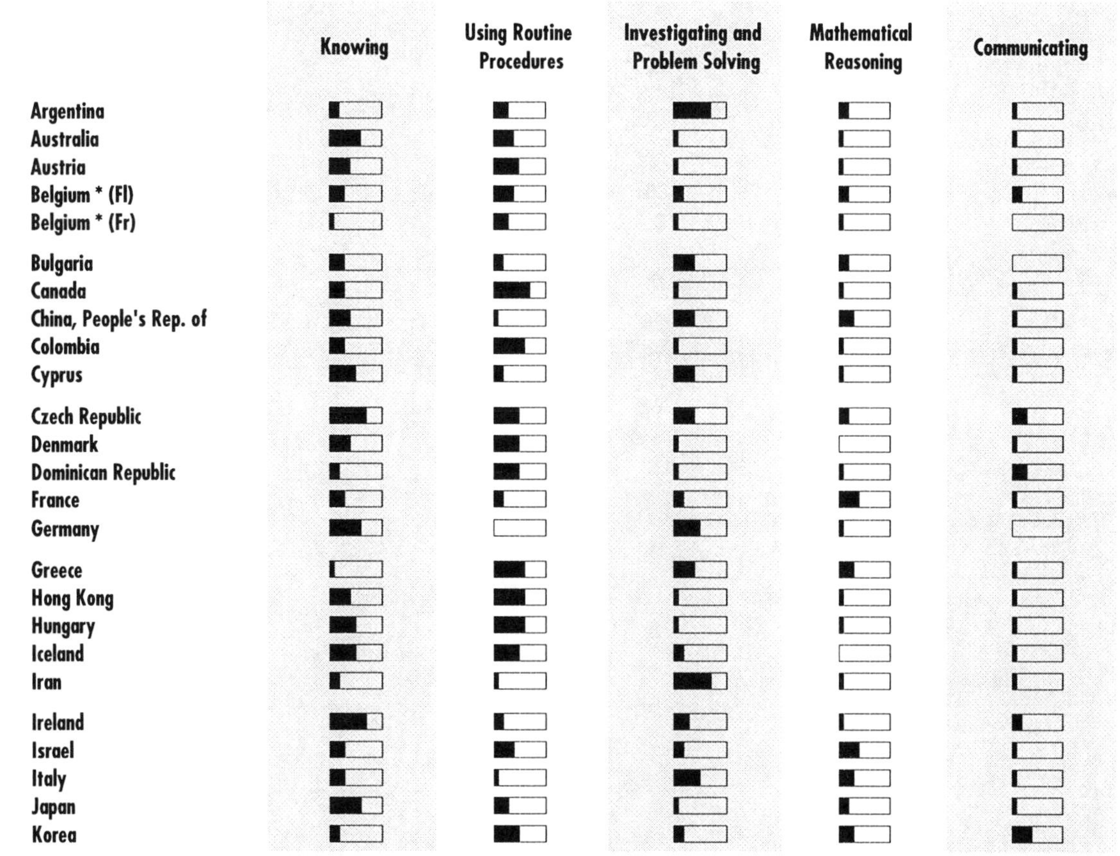

Knowing | Using Routine Procedures | Investigating and Problem Solving | Mathematical Reasoning | Communicating

Argentina
Australia
Austria
Belgium * (Fl)
Belgium * (Fr)

Bulgaria
Canada
China, People's Rep. of
Colombia
Cyprus

Czech Republic
Denmark
Dominican Republic
France
Germany

Greece
Hong Kong
Hungary
Iceland
Iran

Ireland
Israel
Italy
Japan
Korea

* The national Research Coordinators of Belgium have collected data only from curriculum guides. Due to the great level of detail of the guides, and their extensive use, data from these are compared in this display with the textbook data supplied from all other countries.

** Netherlands' sample did not meet the 50% market coverage criterion for Populations 1 and 2.

Note: Percentages across all categories do not sum to 100% because of the multiple codings associated with some blocks.

Missing data: Latvia, Lithuania

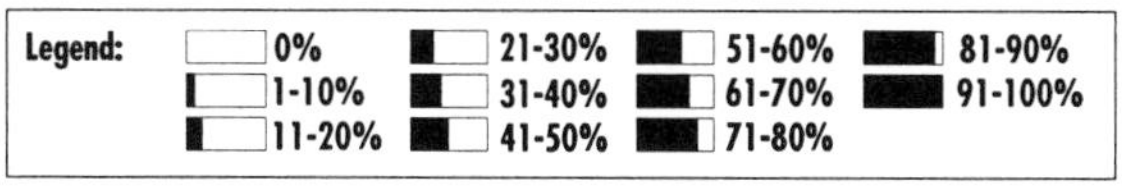

Legend: 0% | 1-10% | 11-20% | 21-30% | 31-40% | 41-50% | 51-60% | 61-70% | 71-80% | 81-90% | 91-100%

Exhibit 26. (cont'd)

Textbook space devoted to major Population 2 mathematics performance expectations.

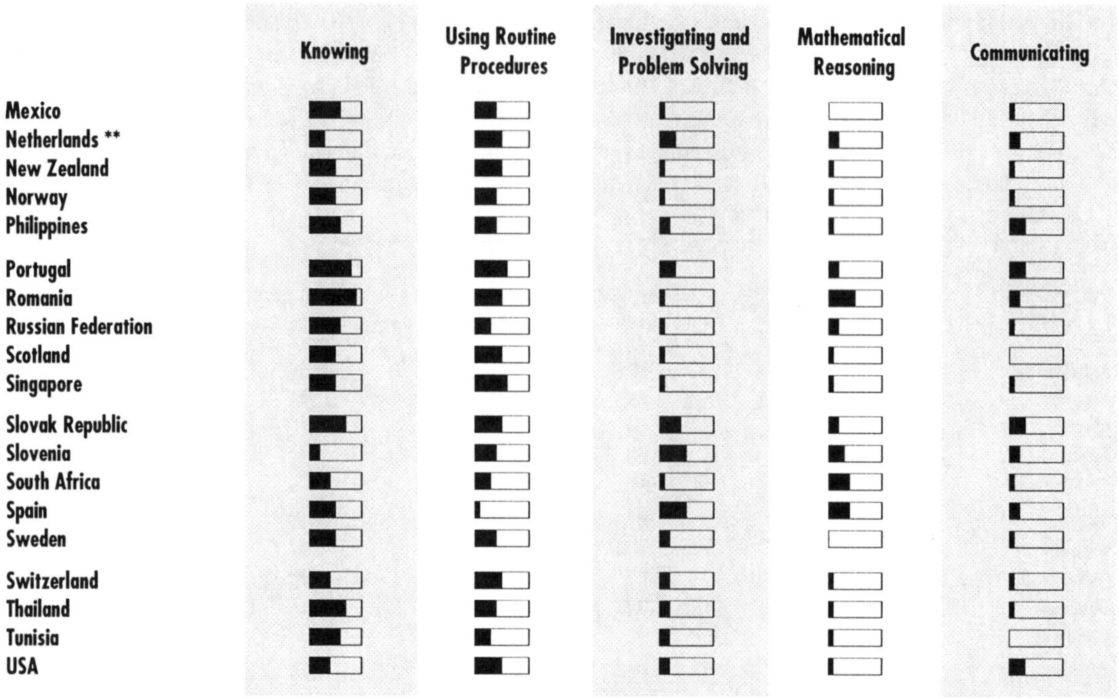

	Knowing	Using Routine Procedures	Investigating and Problem Solving	Mathematical Reasoning	Communicating
Mexico					
Netherlands **					
New Zealand					
Norway					
Philippines					
Portugal					
Romania					
Russian Federation					
Scotland					
Singapore					
Slovak Republic					
Slovenia					
South Africa					
Spain					
Sweden					
Switzerland					
Thailand					
Tunisia					
USA					

* The national Research Coordinators of Belgium have collected data only from curriculum guides. Due to the great level of detail of the guides, and their extensive use, data from these are compared in this display with the textbook data supplied from all other countries.

** Netherlands' sample did not meet the 50% market coverage criterion for Populations 1 and 2.

Note: Percentages across all categories do not sum to 100% because of the multiple codings associated with some blocks.

Missing data: Latvia, Lithuania

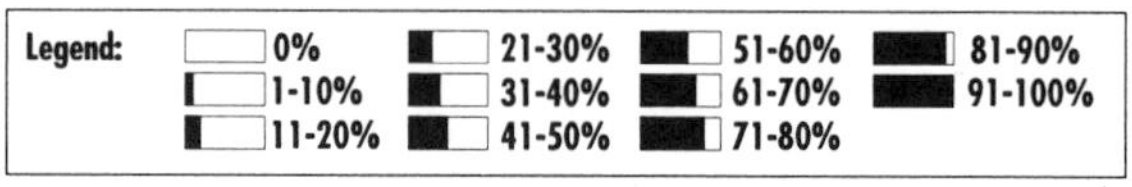

Legend: 0% | 1-10% | 11-20% | 21-30% | 31-40% | 41-50% | 51-60% | 61-70% | 71-80% | 81-90% | 91-100%

Textbooks have been affected to some extent by mathematics education reform recommendations. However, this is primarily through including additional topics without sacrificing more traditional content. It does not generally come through including more demanding, complex, and integrated performance demands. The overall character of U.S. mathematics and science textbooks seems best summarized by saying they are inclusive, largely unfocused, and pay some attention to demanding more of students. However, they still emphasize simpler, less demanding "basic" tasks long a part of traditional U.S. science and mathematics textbooks.

Why is this true? The role played by market mechanisms in shaping textbook designs and revisions offers one possible answer; although, that answer remains conjectural and, while consistent with it, goes beyond the data presented. Commercial publishers must sell U.S. mathematics and science textbooks in changing markets among conflicting demands. Official science and mathematics curricula vary among states and districts, are unfocused, and often seek quite different topics and tasks (exercises and problems). Over 35 states have textbook adoption policies. Unless a textbook meets the criteria stated through these policies or by adoption committees, that textbook cannot even compete for a market share in the states with those adoption policies. In many cases, states approve lists of acceptable textbooks. This simply moves adoption decisions to the more potentially conflict-filled arena of local school and school district adoptions.

U.S. textbook publishers face varied, often conflicting, demands for what should be in mathematics and science textbooks. Such demands range from emphases on more traditional, "basics" materials to including material supporting the agendas of current reforms. These conflicting demands structure the market in which commercial textbook publishing and sales must compete for survival. What is a sensible publishing strategy in this context of a market-based, conflict-filled sales arena? Certainly one key element of any appropriate strategy is to be inclusive, to have something for everybody.

These varied, possibly conflicting direct demands are complemented by the need to provide for successful student performance on common standardized tests. These include state assessments, the National Assessment of Educational Progress (NAEP) tests, commercially produced and locally mandated standardized tests, and so on. In today's educational environment, state and local control over assessment tools and standards for evaluating attainments puts additional pressure on textbook publishers. Many states control widespread use of assessment instruments. In other cases (e.g., California) local districts choose their own assessment instruments by legal mandate. Despite a seeming sameness about most standardized tests, there are differences in content emphases and student performance demands. These differences are enough to provide textbook publishers with yet another set of demands to reconcile in producing and finding a market for their science and mathematics textbooks.

Most schools and teachers make selective use of textbook contents and rarely, if ever, cover all of a textbook's content. Since textbooks are used selectively, publishers are free to provide the needed materials for any common and well-articulated market demand. They can reasonably expect that those who adopt and buy a particular textbook will feel free to use only the contents that suit their purposes. It is far easier to skip unwanted material present in a textbook than to supplement a limited textbook with desired additional material. In a market environment of splintered curricular visions, it is only sound business that U.S. mathematics and science textbooks embody "cautious visions," in particular, a broadly inclusive approach. Unfortunately, the result is large textbooks covering many topics but comparatively shallowly. Even with the largest textbooks in the world space is still limited.

This analysis of cautious visions in U.S. market-driven textbook publishing among conflicting demands needs amplification. One continuing debate concerns the circular relationships between textbooks and markets. Do markets determine the form and content of textbooks? Does what is available in textbooks largely determine what is demanded and, in turn, help shape the textbooks available? The processes shaping textbooks and textbook markets are likely cyclical and iterative. There seems to be no simple resolution of what drives the processes involved.

In this context, textbook producers can hardly be blamed for their cautious, inclusive approaches to the goals and intentions of U.S. mathematics and science education. Given a widely adopted consensus of what is important and central in school mathematics and science, textbooks supporting that vision would soon be in circulation. However, if a clear, coherent vision of the important existed and was shared by virtually all textbook publishers, it is likely that the resulting materials could soon lead to wide official adoption reflecting that coherent vision. In the current U.S. climate of unfocused curricula and cautious, inclusive textbook strategies, we flounder for lack of a shared, coherent vision.

Current reform recommendations in mathematics and science education provide coherent, intellectually satisfying, and well-conceived visions of what should be. Efforts supported by the National Science Foundation (NSF) and others have and continue to support the development of materials designed to fit these reform visions. Unfortunately, hard work, time, and supporting educational policies are needed to allow these visions to be widely shared. While the professional organizations that provided these recommendations work actively to increase the number who share their visions, other participants cannot be absolved from a role in making change occur. If the relationship of textbooks and curricula is cyclical and iterative, textbook publishers share at least some responsibility in fostering change.

Chapter 3

U.S. Teachers: Responding to Splintered Visions

Some facts about teachers and their work do not need the special data of TIMSS to support what has been established repeatedly. Most U.S. teachers are serious professionals. They face a demanding workplace. Many, often conflicting, demands are placed on their time. Teachers rarely have adequate time to think, choose, consult with each other, and make decisions as professionals typically do. While hard working, they are often seriously underprepared in the disciplines of mathematics and the sciences, especially for the demands of new and changing curricula.

As our data demonstrate, U.S. mathematics and science teachers also are asked to teach unfocused curricula stemming from the conflicting opinions and lack of clarity about what should be done. Their main resource in doing so is often inclusive textbooks. These texts provide needed resources but limited guidance and focus. Teachers also must pursue their goals and shape instruction within the real world of their workplace and work day. Faced with these demands, what choices do U.S. science and mathematics teachers make?

What Do U.S. Teachers Do When Asked to Teach Many Topics?

When we ask them to teach more science or mathematics topics than teachers in most other countries, U.S. teachers do what we ask them to do (see Exhibit 27 and Exhibit 28)*. Not surprisingly, U.S. teachers typically report teaching more topics at every percentile of the distrib-

*The results of the Teacher Questionnaire for Population 2 are based on the responses of a sample of 7th and 8th grade teachers within each of the three countries. Thus, as with any sampling-based research, these responses can only be used as an estimate of the responses of the entire population of 7th and 8th grade teachers in each country. In order to increase the precision of this estimate, teachers' responses need to be weighted in relation to the proportion of 7th and 8th grade teachers they represent, and errors of these estimates need to be calculated. The TIMSS sampling design for Population 2 was based on the numbers of 7th and 8th grade *students* in each country and not the numbers of teachers. As a result, it is difficult to estimate the exact relationship between the sample teacher population and the true teacher population with the degree of precision desired for this report. Teacher responses, therefore, need to be interpreted in light of the population of students to which they pertain. However, to help the reader distinguish results based on the Teacher Questionnaire from the results of student surveys, we have chosen to refer to proportions of teachers when discussing exhibits based on teacher data. While these may actually represent quite accurate estimates of teacher responses, they should be interpreted with one of the following two caveats. The results of exhibits that present means should be interpreted as indicating the teacher characteristic to which the typical child is exposed. For example, Exhibits 29 and 30 show an average over teachers' responses on items asking them to indicate the number of

Exhibit 27.

Number of topics covered by grade 8 mathematics teachers.

U.S. grade 8 mathematics teachers typically teach far more topics than their counterparts in Germany and, especially, Japan. [Each box portrays a sample of teachers for either Germany, Japan, or the U.S. Each row within a box corresponds to one sampled teacher. Each column within a box corresponds to one topic area that teachers were asked whether they taught during the sampled school year. A dash indicates the teacher represented by that row indicated that she or he taught the topic represented by that column. A blank indicates the teacher reported the topic as not taught. We ordered the rows from teachers reporting teaching more topics (upper rows) to those reporting teaching fewer (lower rows). We arranged the columns so that topics more commonly reported as taught are to the left and less commonly reported to the right. The solid line drawn across each box indicates how many topics each teacher taught. Farther to the right means more topics taught. A line that moves more slowly to the left as it moves from top to bottom indicates teachers typically teaching more topics than a line that moves more quickly to the left. The sense of a more completely "shaded" or filled box compared to other boxes also indicates that more topics typically reported as taught.]

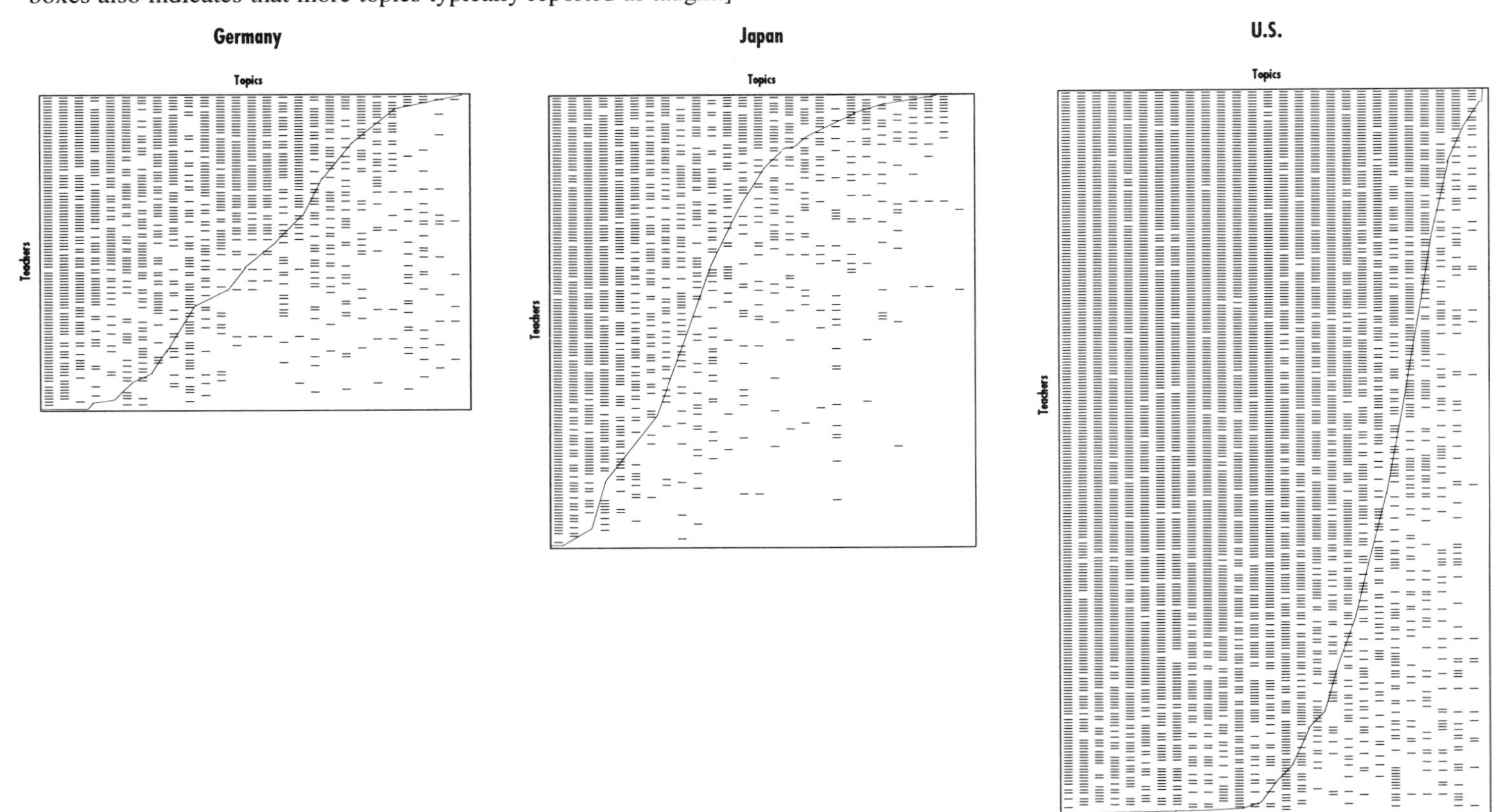

Exhibit 28.

Number of topics covered by grade 8 science teachers.

U.S. grade 8 science teachers typically teach far more topics than their counterparts in Japan and, especially, Germany. [Each box portrays a sample of teachers for either Germany, Japan, or the U.S. Each row within a box corresponds to one sampled teacher. Each column within a box corresponds to one topic area which teachers were asked whether they taught during the sampled school year. A dash indicates the teacher represented by that row indicated that she or he taught the topic represented by that column. A blank indicates the teacher reported the topic as not taught. We ordered the rows from teachers reporting teaching more topics (upper rows) to those reporting teaching fewer (lower rows). We arranged the columns so that topics more commonly reported as taught are to the left and less commonly reported to the right. The solid line drawn across each box indicates how many topics each teacher taught. Farther to the right means more topics taught. A line that moves more slowly to the left as it moves from top to bottom indicates teachers typically teaching more topics than a line that moves more quickly to the left. The sense of a more completely "shaded" or filled box compared to other boxes also indicates that more topics typically reported as taught.]

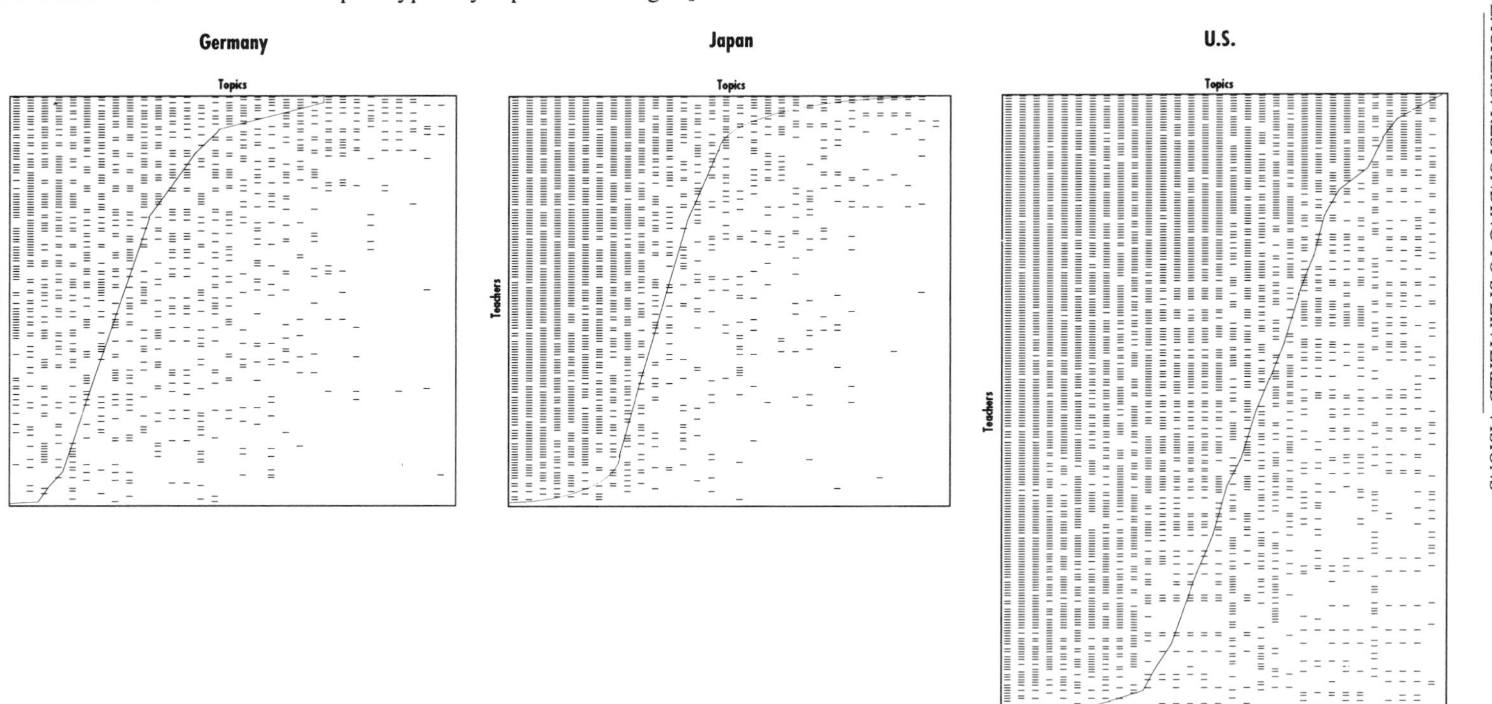

ution of sampled teachers than do German and Japanese teachers. U.S. teachers report teaching not only large numbers of topics, but also larger numbers of topics than their counterparts in other countries.

How Much Time Do U.S. Teachers Typically Spend on a Topic?

Instructional time for science or mathematics is limited in any school year. Since teachers work with many topics, this limited time suggests that they spend little time on average for most topics. Our data support this. (Exhibit 29 and Exhibit 30 display these data for grade 8* for the U.S., Germany, and Japan.) Grade 8 teachers spent more than 20 instructional periods— that is at least about four school weeks — on few topics in the U.S., Germany, and Japan. In Japan, at grade 8 four of the topics extensively emphasized by textbooks also received extensive attention by teachers — two geometry topics and the two algebra topics. In Germany, textbooks devoted extensive space to some geometry and algebra. Teachers shared the emphasis on algebra but emphasized 'fractions and decimals' rather than geometry. In the U.S. both regular and pre-algebra teachers emphasized only one topic, 'fractions and decimals.' U.S. grade 8 algebra teachers also emphasized 'fractions and decimals' as well as 'equations and formulas' (a part of algebra). In science, only earth and life science teachers emphasized any topics ('earth features,' the structure of living things, life systems, and genetics). Teachers from Germany and Japan emphasized two topics each.

Other than the exceptions noted above, teachers in the U.S., Germany, or Japan did not give extensive attention to any other topics. From the choices available in the questionnaire, the most commonly reported amount of time allocated by teachers was 1 to 13 periods.

U.S. grade 8 teachers reported devoting some attention to most all science or mathematics topics about which we asked. U.S. grade 8 mathematics teachers and grade 8 earth and general science teachers report teaching all topics. This is direct evidence that this sample of teachers covered a little of everything. They devoted extensive time (more than 13 periods) to only a few topics.

These data supplement the earlier data that U.S. teachers covered more topics than Germany and Japan. This "breadth" was achieved at the price of "depth." Teachers could cover many topics only by limited instructional time for virtually every topic. This further indicates at most modest coverage of each topic.

---

hours they spend teaching certain topics. This exhibit can be interpreted as the number of hours a typical student's teacher spent teaching each topic. On the other hand, results of exhibits that present proportions or percentages should be interpreted in light of the proportion or percent of students whose teachers behave in the ways presented. For example, Exhibit 35 shows the percentage of teachers who indicate familiarity with various educational documents. This exhibit can be interpreted as the percentage of students whose teachers are familiar with each document. Therefore, although the exhibits are worded as teacher characteristics, it is more proper to interpret the responses in relation to students. Standard errors of all data based on the teacher survey are presented in Appendix C.

*While not true in all TIMSS countries, Population 2 is grades 7 and 8 in the U.S., Japan, and Germany and we can refer to the grade levels rather than the population number.

Exhibit 29.

Number of periods grade 8 mathematics teachers covered specific topics.

Grade 8 mathematics teachers surveyed in Germany, Japan, and the U.S. indicated for the most part that they taught a few periods of almost every topic listed. [We asked teachers to report within specified ranges the number of periods that they had devoted to specific mathematics topics. We surveyed teachers from both grades of Population 2. We obtained the topics used for our survey by combining more specific topics from the TIMSS' mathematics framework to yield a list of familiar topics broadly appropriate to these grade levels. The instrument given teachers to gather these data provided brief descriptions of each topic. We use four different icons here to show which ranges of intended content coverage grade 8 teachers reported on average for each of the countries.]

	Average Number of Periods				
	Grade 8				
	Germany	Japan	U.S. Regular	U.S. Pre-Algebra	U.S. Algebra
Whole Numbers	❍	❍	❍	❍	❍
Fractions & Decimals	◑	❍	●	●	◑
Other Numbers and Number Concepts	❍	❍	❍	❍	❍
Number Theory and Counting	❍	○	❍	❍	❍
Estimation and Number Sense	❍	❍	❍	❍	❍
Measurement Units	❍	❍	❍	❍	❍
Perimeter, Area and Volume	❍	❍	❍	❍	❍
Measurement Estimation and Error	❍	❍	❍	❍	❍
2-D Geometry	❍	◑	❍	❍	❍
Geometry Transformations	❍	❍	❍	❍	❍
Congruence and Similarity	❍	●	❍	❍	❍
3-D Geometry, Vectors, Constructions	❍	❍	❍	❍	❍
Proportionality Concepts	❍	❍	❍	❍	❍
Proportionality Problems	❍	❍	❍	❍	❍
Slope and Trigonometry	❍	❍	○	❍	❍
Linear Interpolation and Extrapolation	○	❍	○	○	❍
Patterns, Relations, and Functions	❍	◑	❍	❍	❍
Equations and Formulas	●	●	❍	❍	◑
Data Representation and Analysis	❍	❍	❍	❍	❍
Uncertainty and Probability	❍	○	❍	❍	❍
Sets and Logic	❍	○	❍	❍	❍
Other Content	❍	❍	❍	❍	❍

Legend:
○ = <1 period
❍ = 1 - 13 periods
◑ = 14 -19 periods
● = >19 periods

Exhibit 30.

Number of periods grade 8 science teachers covered specific topics.

Grade 8 science teachers surveyed in Germany, Japan, and the U.S. indicated that they taught at least a few periods of most every topic listed. Grade 8 physical and general science teachers indicated that there was no topic on which they spent more than 13 instructional periods. A few topics received more extensive coverage in Japan and Germany, as well as by U.S. earth and life science teachers. [We asked teachers to report within specified ranges the number of periods that they had devoted to specific science topics. We surveyed teachers from both grades of Population 2. We obtained the topics used for our survey by combining more specific topics from the TIMSS' science framework to yield a list of familiar topics broadly appropriate to these grade levels. The instrument given teachers to gather these data provided brief descriptions of each topic. We use four different icons here to show which ranges of intended content coverage grade 8 teachers reported on average for each of the countries.]

	Average Number of Periods					
	Grade 8					
	Germany	Japan	U.S. Earth	U.S. Life	U.S. Physical	U.S. General
Earth Features	◔	◔	◑	◔	◔	◔
Earth Processes	◔	◔	◔	◔	○	◔
Earth in the Universe	◔	◔	◔	◔	◔	◔
Div., Org. & Struc. of Living Things	◑	◑	◔	◑	○	◔
Life Processes & Systems	◑	◑	◔	◑	○	◔
Life Spirals, Genetic Continuity & Div.	◔	◔	◔	◑	◔	◔
Interactions of Living Things	◔	◔	◔	◔	○	◔
Human Biology & Health	◔	◔	◔	◔	○	◔
Matter	◔	◔	◔	◔	◔	◔
Structure of Matter	◔	◔	◔	◔	◔	◔
Energy & Physical Processes	◔	◔	◔	◔	◔	◔
Physical Transformations	◔	◔	◔	◔	◔	◔
Chemical Transformations	◔	◔	◔	◔	◔	◔
Forces & Motion	◔	○	◔	○	◔	◔
Science, Technology & Society	◔	○	◔	◔	◔	◔
History of Science & Technology	◔	◔	◔	◔	◔	◔
Environmental & Resource Issues	◔	◔	◔	◔	◔	◔
Nature of Science	◔	◔	◔	◔	◔	◔

Legend:
○ = <1 period
◔ = 1 - 13 periods
◑ = 14 -19 periods
● = >19 periods

How Much Time Do U.S. Teachers Spend on the Few Topics to Which They Devote the Most Attention?

We can examine the percentage of instructional time teachers devote to the few topics on which they spend the most time. (See Exhibit 31 and Exhibit 32, which compare to the textbook data in Exhibit 23 and Exhibit 24.) From these data clearly U.S. grade 8 mathematics and general science teachers devoted less than half of their time to covering the five most highly covered topics. Teachers appear to have distributed the rest of the year's instructional time among the other topics surveyed without much focus. These data further indicate that teachers distribute instructional time in an inclusive, unfocused manner similar to official U.S. curricula and textbooks.

There are important exceptions to this general finding. U.S. grade 8 science instruction for most students (about 70 percent) took place in courses focused on one area of science — physical, life, or earth science. Both the textbooks for those courses and their teachers' time allocations were much more focused than grade 8 "general" science courses. However, even in these cases, with specifically focused courses and highly focused textbooks, the U.S. teachers reported teaching a little of a wide variety of topics, even those that were not a major part of the area of the sciences on which their courses were focused. A similar case held in mathematics for grade 8 Algebra I courses, which had both more focused texts than other mathematics courses and slightly more focused topic coverage by teachers (although, again, something of almost everything was taught). For both the single area science courses and the Algebra I courses, the teachers indicated less focus than the corresponding textbooks.

The five topics receiving the most textbook space came from the more extensive collection of specific topics in the TIMSS' mathematics and science frameworks. The five topics receiving most teacher attention came from the reduced list created by combining more specific framework topics into categories more easily recognizable by grade 8 teachers. We cannot disaggregate the teacher data to look at more specific topics used for textbooks. However, we can aggregate the textbook topic data into the combinations to which teachers responded.

After we aggregated the textbook data to match the teacher survey content categories, we still find results similar to the previous textbook results. This analysis revealed one new result. Using this reduced, aggregated list of topics for U.S. grade 8 mathematics, the five most covered topics for textbooks and for teachers had three topics in common. These topics were 'fractions and decimals,' 'other numbers and number concepts,' and 'equations and formulas.' The topics most extensively covered by teachers were similar but not identical to the topics most extensively covered by U.S. grade 8 mathematics textbooks. U.S. grade 8 mathematics teachers appear to be using textbooks selectively. Teachers covered some topics at a more leisurely pace but others more briefly. The "pace" of the textbooks matches the "pace" of the teacher only in some cases. Four of five topics were the same in Japanese grade 8 mathematics. U.S. general and composite science teacher coverage matched corresponding textbook coverage only in very limited ways.

Exhibit 31.

The five topics emphasized most by grade 8 mathematics teachers.

U.S. grade 8 mathematics teachers' five most extensively covered topics accounted for less than half of their instructional periods. This is even more so since one of the topics was 'other content,' a combination of smaller topics. This was true both for Algebra I teachers and other grade 8 mathematics teachers. Japanese teachers' five most extensively covered topics included almost 75 percent of instructional periods (65 percent if 'other content' is omitted). U.S. teachers' actual content coverage echoes the inclusive, unfocused character of U.S. official curricula and textbooks in mathematics. [Each bar shows the cumulative percentage of intended class periods, averaged across teachers, devoted to the five topics allocated most time by grade 8 mathematics teachers. We show bars for Japan, Germany, and the U.S. Each bar segment shows the specific TIMSS framework topics (e.g., 'equations and formulas') involved. Corresponding international data are not yet approved for release.]

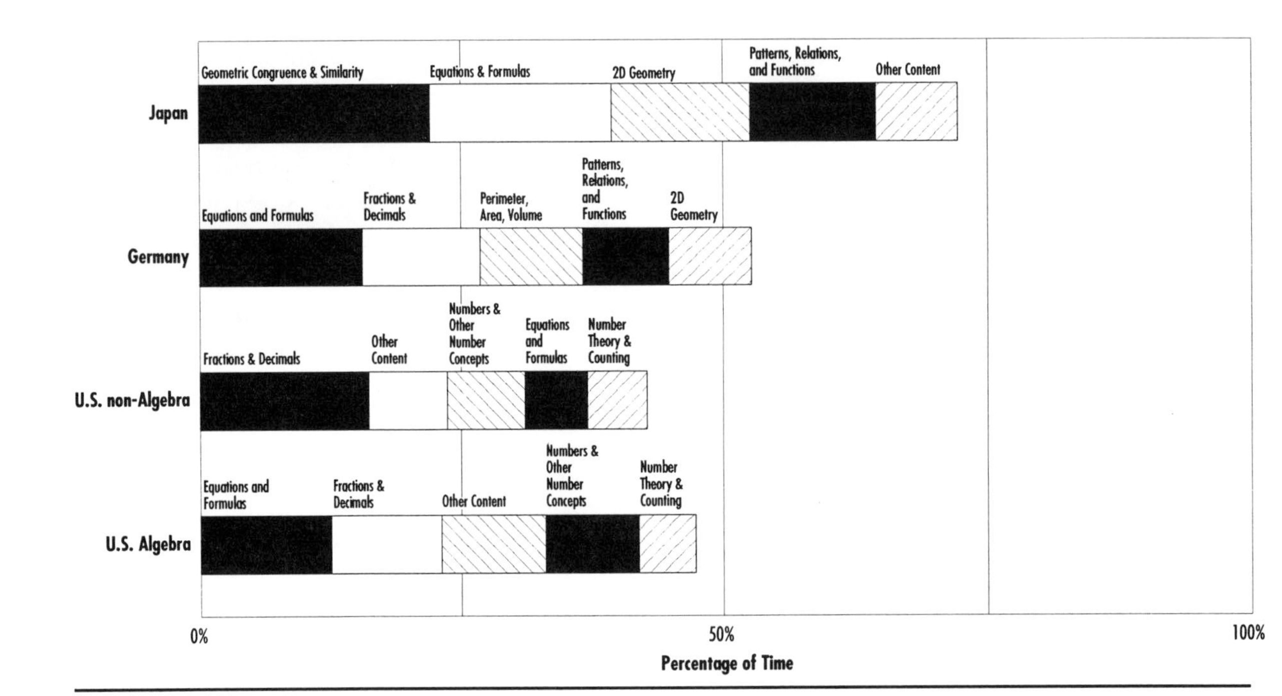

Exhibit 32.

The five topics emphasized most by grade 8 science teachers.

U.S. grade 8 general science teachers' five most extensively covered topics accounted for only about 40 percent of their instructional periods. Japanese teachers' five topics most extensively covered included over 60 percent of instructional periods, and German teachers' five topics about 50 percent. U.S. grade 8 teachers for specialized science courses ('earth science,' etc.) showed more focused coverage. Their five most extensively covered topics included over 50 percent of instructional time. [Each bar shows the cumulative percentage of intended class periods, averaged across teachers, devoted to the five topics allocated most time by grade 8 science teachers. We show bars for Japan, Germany, and the U.S. Each bar segment shows the specific TIMSS framework topics (e.g., 'earth features') involved. Corresponding international data are not yet approved for release.]

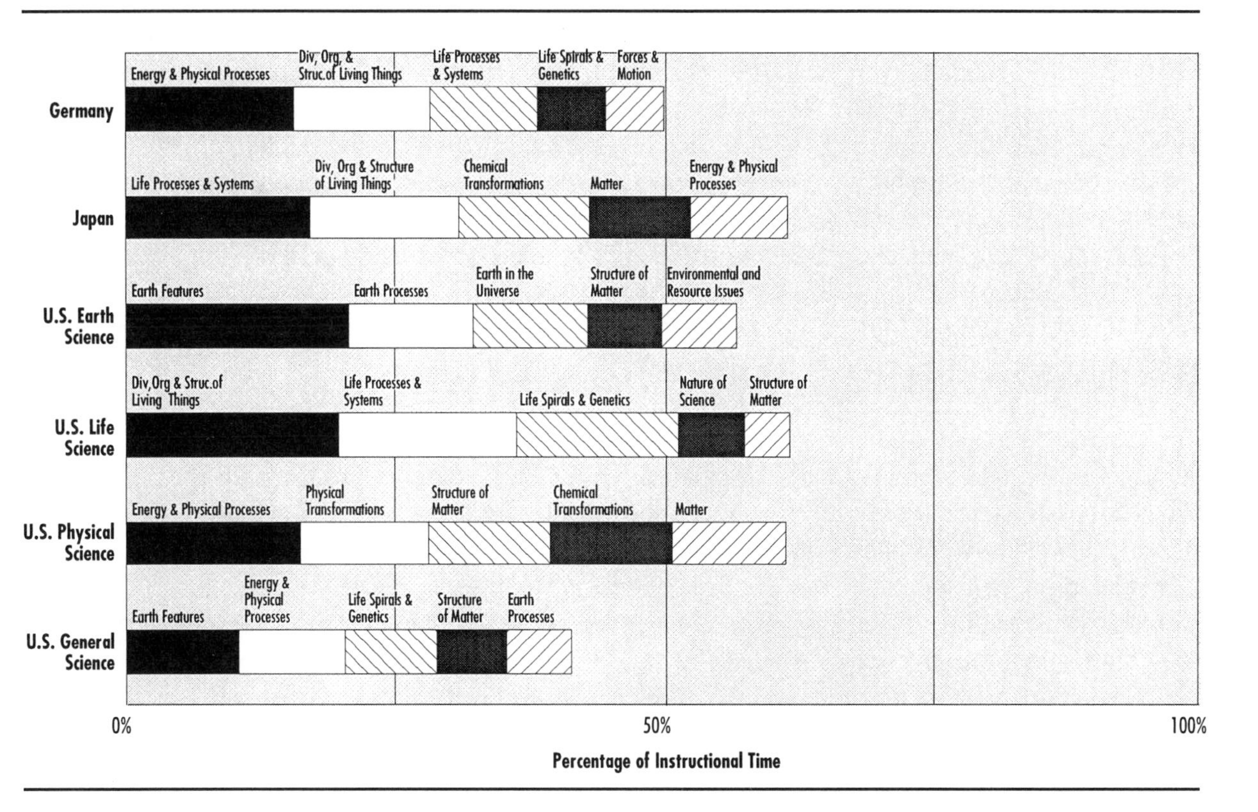

The above findings seem at first evidence that U.S. teachers are less tied to textbooks than commonly thought. Unfortunately, this is only one way of interpreting these results. All teacher choices were contents that matched some content in corresponding textbooks. Only the amounts of teacher instructional time and the comparative textbook space differed. This is not conclusive evidence that teachers supplemented textbooks and omitted topics from them; although, this likely happened in some cases.

These same data may simply indicate that teachers covered selected sections of textbooks at a pace disproportionate to the amount of textbook space allocated to the topic. Teachers can change the pace and emphasis by using more or less in-class discussion. They can also vary how many exercises and questions they use from textbooks (since the books provide more than can be used for most topics). Teachers can supplement text coverage with other materials and activities, but this is only one way for teachers to change the pace and emphases provided in their textbooks. These data certainly imply that teachers used their textbooks *selectively* in many cases. They covered topics at more or less detailed levels at their own discretion or by following local policy guidelines (e.g., local school 'scope and sequence' or 'syllabi' guidelines). Teachers may have also supplemented textbooks in addition to using them selectively, or they may have combined both selectivity and supplementing.

How Do U.S. Teachers Decide What to Cover in Daily Lessons?

U.S. mathematics and science teachers have demanding working conditions. They make their decisions and plans for daily instruction under those conditions. How do they plan how they will use instructional time? Teachers appear to use textbooks selectively, and this selective use requires planning and decision-making. Even if teachers simply vary the detail with which they cover textbook sections and the time spent on them, how do they make these decisions? The data provide little direct insight to answer this question. We are left to search for possibilities sufficient enough to explain what the data do reveal, to reflect a sense of the data, and to further discussion aimed at meaningful change. One relevant notion is that of "satisficing."[12] This idea seems sufficient to explain much teacher behavior, especially in their coverage of topics and in their use of textbooks.

"Satisficing" suggests that teachers use an approach of "satisfying sufficiently" and making choices that are "good enough" under the circumstances. The notion is that those who "satisfice" make choices that satisfy the system that guides their actions but without trying for the absolute best decision. They make a decision "good enough" to satisfy their own sense of responsibility. The order in which options are considered is also important to the notion of "satisficing." Those who "satisfice" will typically stop with the first decision among alternatives considered that seems good enough to them. Thus, they make decisions relative to the order in which they consider alternatives.

U.S. teachers' instructional decisions must be made quickly under demanding circumstances. Teachers who satisfice using focused curricula or textbooks that present prioritized alternatives may make more optimal decisions, even when seeking only "good enough" alternatives. Unfocused curricula or inclusive textbooks fail to aid teachers in recognizing or setting priorities and are especially likely to affect the decisions of teachers who satisfice.

U.S. teachers who satisfice are guided by their sense of responsibility. That sense shapes their criteria of what is "good enough." They do what seems "good enough" among the choices realistically available to them for the children they teach. Teachers seem to regard textbooks and curricula as sanctions indicating what is acceptable. If a curriculum is unfocused and does not set priorities, or if textbooks are inclusive and do not organize priorities, teachers who satisfice will likely decide what to do on the basis of what they encounter and, thus, enact decisions that reflect the unfocused, inclusive natures of the textbooks and curricula. That is, it seems likely that teachers often settle for textbook-based default choices because these are the first alternatives they consider, they help sequence alternatives and choices, and they are considered to contain officially sanctioned things "good for the children."

Of course, other factors affect teachers' decisions. Many teachers rely on their own teaching experience, on what they have done in similar situations in the past, or on their own experience with the same textbook and its successful or unsuccessful use. They rely on what collegial support they can obtain. In the end, they rely on what feels comfortable and appeals to them. Unfortunately, there are indications that U.S. teachers are weaker in subject-matter preparation and knowledge than teachers in other countries.

The satisficing conjecture is not meant to make teachers villains in the current state of U.S. mathematics and science education. Some teachers are burned out; some have ceased to invest themselves; some try their hardest every day. Many are underprepared in subject-matter knowledge. However, some teachers, rise above satisficing to seek what is truly best. They seek and find what is deep and fundamental, and what can make differences in the lives of receptive children. These teachers consider a wider array of alternatives until they find a choice better than the merely "good enough." They do this through sacrificial expenditures of time, through insight into their subject-matter, and through experience sifted by reflection. These teachers likely have deeply held personal values that lead them to stronger, more demanding conceptions of what is "best for the children."

Currently, we encounter almost by chance those who rise above satisficing. Policy-makers must seriously consider how to increase the number of teachers who do more than satisfice and the number of occasions on which they do so. Better preparation and subject-matter knowledge may make more instructional alternatives more readily come to mind for more teachers. Essentially, we must determine what will change teachers' conceptions of what is best, what will increase the alternatives considered, and what will provide them the opportunity to do more than the "good enough."

It seems likely that fundamental changes are needed. Changes are needed in preparing teachers with thorough knowledge of mathematics or the sciences. Fundamental changes are needed in their working conditions. Changes are needed in the quality of curricula guiding them, and widely accepted curricula are needed with more serious visions of what children deserve. Textbooks are also needed that more effectively support decision-making beyond satisficing but remain commercially viable for their publishers.

Without fundamental changes such as these, we are unlikely to decrease the probability of U.S. mathematics and science teachers finding it necessary to satisfice. Without these supporting changes, the path of content-based reform recommendations is more difficult and, to the extent they rely on information and persuasion, their success may be limited even when these reform efforts offer useful visions of what might and ought to be.

Satisficing is one possible explanation of how teachers make instructional plans and how they interact with textbooks and curricula in doing so. Further, this conjecture is consistent of what we found about teacher topic coverage and textbook use. While we have found it difficult to gather data that directly clarifies the nature of teachers' instructional decision-making and how they use textbooks in planning instruction, other research reveals clearly that many (perhaps most) mathematics and, to a lesser extent, science teachers rely heavily on their textbooks. However, not all do so and most do not do so completely, as our data suggest.

Developmental research supporting instrument development for TIMSS (reported elsewhere[13]) sheds some light on teachers' use of textbooks. We collected these data through observations and interviews with science and mathematics teachers in six countries. Different factors are relevant in different countries. In France, Population 2, we found mathematics teachers to have subject-matter training sufficiently extensive that they considered themselves content specialists. They correspondingly more often relied on their own training and judgment rather than on a particular textbook although they still made use of textbooks to support instruction. In France, Japan, and at least parts of Switzerland, teachers functioned in a more focused system that gave teachers much more detailed guidance. This seems both to have affected the instructional alternatives they considered and their standards of what was good for their students.

The TIMSS survey data does indicate that U.S. teachers use textbooks as part of their daily instruction to about the same degree as their Japanese and German counterparts. Mathematics teachers from all three countries indicated that they based about 60 percent of their weekly teaching time directly on the textbook (closer to 70 percent in Japan). For science teachers the amount was about 50 percent (closer to 60 percent in Japan).

Textbooks obviously play an important role in structuring instruction. This is likely to be even more true in demanding instructional situations. This includes situations that emphasize accountability and in which the costs for "indefensible" teacher "independence" are high. It includes situations in which teachers are not so strongly prepared in their subject-matters that they can consider themselves specialists. It certainly includes situations in which teachers' decisions are not supported by focused curricula and textbooks. It also raises an important question: do U.S. teachers even know how to do better than they have typically done?

Is This the Best Our Teachers Can Do?

Teachers currently work, as hypothesized in the previous section, under conditions that make satisficing a reasonable strategy for psychic survival. Are they doing the best they know to do? We can help answer this by examining some of the beliefs of U.S. teachers, how informed they are, and how willing they are to face hard work.

Do teachers' beliefs suggest they are capable of more than what they usually do? We surveyed teachers' beliefs about pedagogical strategies most appropriate to specific teaching situations. We asked two questions of both mathematics and science teachers. Each pair described specific subject-matter contents and distinct pedagogical situations. Within each situation, we provided teachers with a list of pedagogical approaches and asked them to choose, on the basis of their beliefs, which would be better. (Exhibit 33 shows one of the items used with Population 2 mathematics teachers.)

How did teachers respond to these items? We classified the specific responses among which teachers chose into four fundamentally different pedagogical strategies. Very few U.S. teachers indicated that sticking to their textbook's approach was the best strategy in either situation (see Exhibit 34). More Japanese teachers indicated a preference for the textbook's strategy when introducing a new topic. Obviously they were dealing with Japanese textbooks that reflected both a different status for teachers and a different strategy for supporting strongly supervised approaches to topics.

Most grade 8 U.S. teachers chose a deductive approach for introducing a topic and an inductive approach for correcting a misconception. We can question the wisdom of the deductive approach. It may indicate a more traditional approach of giving rules and examples. However, unquestionably the type of strategy preferred varied with the pedagogical situation. Textbooks did not dominate U.S. teachers' strategies in either case. Given a chance to indicate what they considered pedagogically best, the U.S. teachers indicated belief in using varied, situationally appropriate strategies.

In science, more than half of both U.S. and Japanese teachers preferred an approach based on conducting a relevant investigation and collecting data to both pedagogical situations. Both sets of teachers also rejected following the textbook. The U.S. teachers' pedagogical preferences strongly mirror a major emphasis of the National Science Education *Standards*. This contrasts with typical U.S. grade 8 science instruction. Words, pictures, talk, and the textbook often dominate this actual instruction. It is less often based on empirical investigation. Again, the preferred was not the practiced.

Exhibit 33.

Sample item on pedagogical beliefs for Population 2 mathematics teachers.

This exhibit shows one item used to gather information on teachers' beliefs about more and less effective pedagogical approaches. This particular item was for Population 2 mathematics teachers. It deals with introducing a new topic and with a particular mathematics topic. A parallel item used counterparts of each strategy listed but different mathematics content and sought teachers' beliefs about correcting a widely held student misconception.

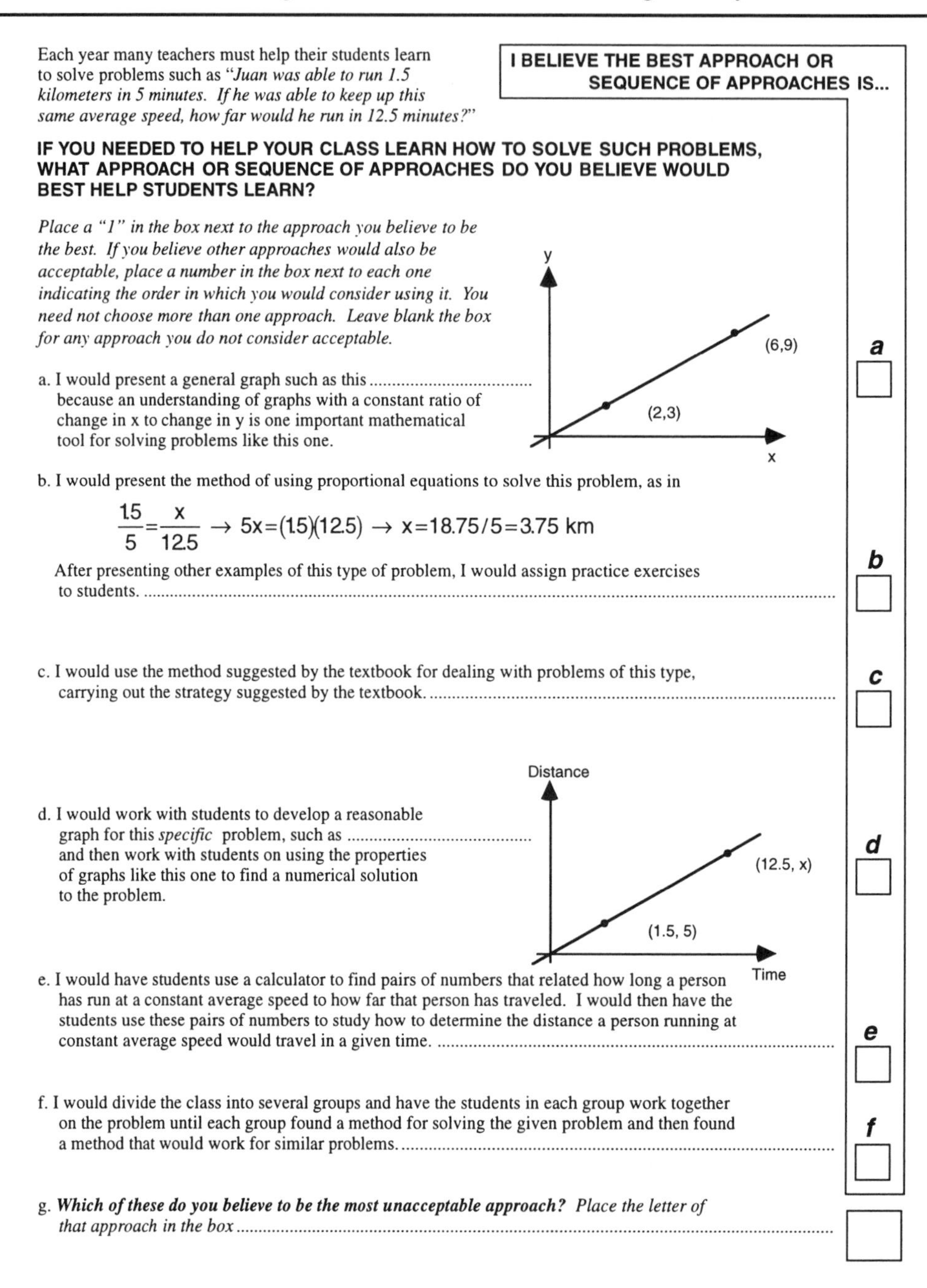

Each year many teachers must help their students learn to solve problems such as *"Juan was able to run 1.5 kilometers in 5 minutes. If he was able to keep up this same average speed, how far would he run in 12.5 minutes?"*

I BELIEVE THE BEST APPROACH OR SEQUENCE OF APPROACHES IS...

IF YOU NEEDED TO HELP YOUR CLASS LEARN HOW TO SOLVE SUCH PROBLEMS, WHAT APPROACH OR SEQUENCE OF APPROACHES DO YOU BELIEVE WOULD BEST HELP STUDENTS LEARN?

Place a "1" in the box next to the approach you believe to be the best. If you believe other approaches would also be acceptable, place a number in the box next to each one indicating the order in which you would consider using it. You need not choose more than one approach. Leave blank the box for any approach you do not consider acceptable.

a. I would present a general graph such as this .... because an understanding of graphs with a constant ratio of change in x to change in y is one important mathematical tool for solving problems like this one. **a** ☐

b. I would present the method of using proportional equations to solve this problem, as in

$$\frac{1.5}{5}=\frac{x}{12.5} \rightarrow 5x=(1.5)(12.5) \rightarrow x=18.75/5=3.75 \text{ km}$$

After presenting other examples of this type of problem, I would assign practice exercises to students. .... **b** ☐

c. I would use the method suggested by the textbook for dealing with problems of this type, carrying out the strategy suggested by the textbook. .... **c** ☐

d. I would work with students to develop a reasonable graph for this *specific* problem, such as .... and then work with students on using the properties of graphs like this one to find a numerical solution to the problem. **d** ☐

e. I would have students use a calculator to find pairs of numbers that related how long a person has run at a constant average speed to how far that person has traveled. I would then have the students use these pairs of numbers to study how to determine the distance a person running at constant average speed would travel in a given time. .... **e** ☐

f. I would divide the class into several groups and have the students in each group work together on the problem until each group found a method for solving the given problem and then found a method that would work for similar problems. .... **f** ☐

g. ***Which of these do you believe to be the most unacceptable approach?*** *Place the letter of that approach in the box* .... ☐

Exhibit 34.

Pedagogical approaches grade 8 mathematics teachers chose in two pedagogical situations.

More than half of U.S. grade 8 mathematics teachers chose a deductive approach for introducing a new topic. In contrast, almost 75 percent chose an inductive approach for correcting a misconception. Very few chose a text-dominated approach. Japanese teachers were more varied in the strategies they thought best. [These data indicate the percentage of grade 8 mathematics teachers who gave as their first choice an option from one of four pedagogical approaches underlying a more specific list of options. Exhibit 33 shows the full text of Item 1. Item 2 was a parallel item dealing with correcting a widely held student misconception.]

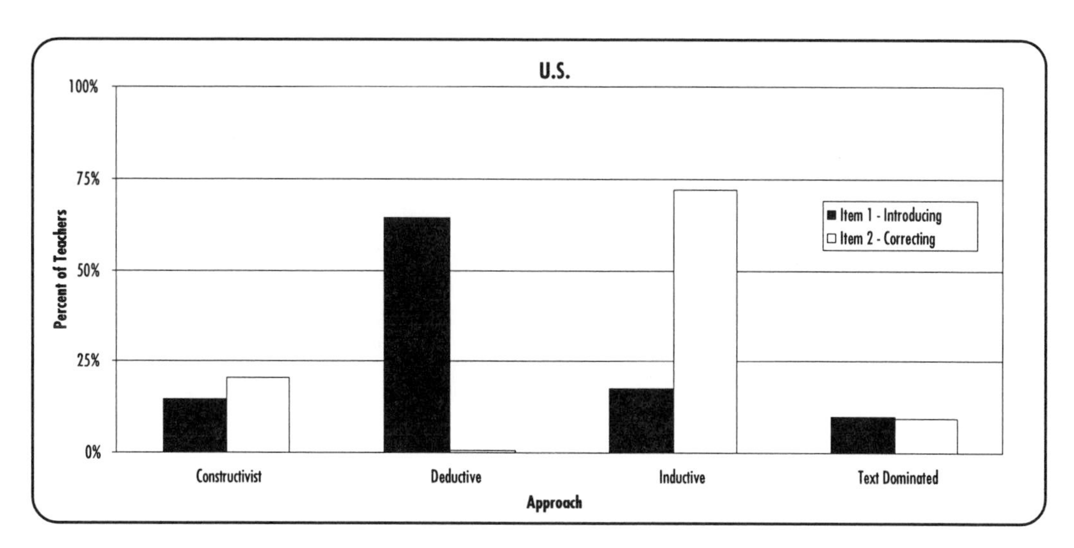

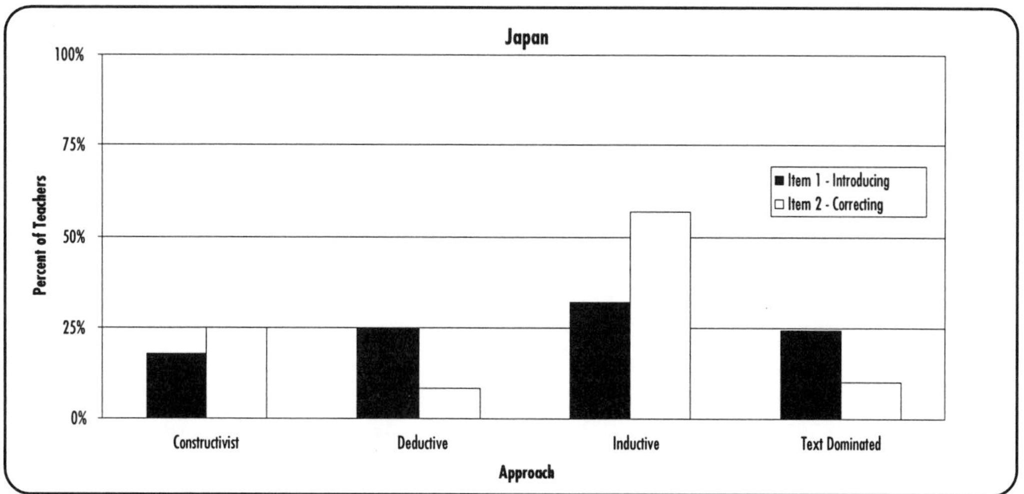

Clearly, the commonalities of typical pedagogical practice by U.S. teachers is not — or not solely — due to their unawareness of other options. They are aware of them and believe in the effectiveness of at least some. Teachers responded with a more varied, adaptive pedagogical style when asked their beliefs. What they actually do varies less and textbooks more often dominate it. Some of the survey response may be due to "telling the questioner what he or she wants to hear." However, such pronounced and consistent differences are not likely due entirely to attempts to please. Real beliefs were to some extent reflected in the responses. To the extent that they were, we must conclude that the typical teaching provided by U.S. mathematics and science teachers is not necessarily the best of which they are capable.

Are U.S. teachers well informed about documents and directions that should make their teaching more effective? Are typical classroom teachers aware of the voices and visions with relevant information that might affect their instructional practices? Wide familiarity with the dominant mathematics reform report was claimed by more than 80 percent of the mathematics teachers (see Exhibit 35). It is inappropriate to compare the mathematics and science results since the science reform document referenced was far more recent. The data suggest that, with time, teachers become familiar with the dominant voices seeking reform in their subject-matter area.

Over 60 percent of the grade 8 teachers reported familiarity with both state and district curriculum guides (over 75 percent for district guides). Official documents meant to guide practices within states and districts apparently are penetrating as widely as state and local education agencies likely hope.

Overall we must conclude that the majority of teachers are familiar with one or more documents intended to inform their instruction and increase its effectiveness. Familiarity with national reform documents, with time, suggests an element of professionalism in teachers and that relevant information is reaching them. Most teachers' current practices appear not to be shaped by the lack of information about state curriculum guides and other documents. Rather, that information appears insufficient to overcome satisficing and other effects of teacher working conditions. Providing more and better information to teachers is not likely to be enough to produce needed change. Other changes are likely more fundamental and essential for improving classroom instruction.

Are U.S. teachers willing to work hard? Are U.S. teachers willing to work hard? U.S. teachers do work hard. Our data show U.S. science and mathematics teachers are scheduled to work about 30 periods each week (see Exhibit 36). There must be at least modest preparation and grading time for what are likely six classes. There is typically also out-of-class work with individual students and non-teaching school responsibilities. In contrast, German teachers are scheduled to work just over 20 periods per week and Japanese teachers fewer than 20. Clearly the working conditions for German and Japanese teachers are distinctly different from those for

Exhibit 35.

Grade 8 mathematics and science teacher awareness of curriculum-relevant documents.

More than 75 percent of U.S. grade 8 mathematics teachers indicated familiarity with the NCTM *Standards* and over half reported familiarity with their state's and district's official curriculum guide. From 30 to 40 percent reported familiarity with their state assessment specifications. Familiarity with NAEP examination specifications was low (less than 25 percent). Science teachers reported being even less familiar than math teachers with NAEP specifications and much less familiar with the (more recent) science reform report mentioned. They indicated similar familiarity with their state's and district's curriculum guide and state assessment specifications. [We surveyed teachers to determine whether they considered themselves familiar with the contents of relevant curriculum documents. In each case, the documents we asked teachers about were a national reform report (e.g., the NCTM *Standards*), each teacher's official state and district curriculum guide, and the specification of state assessments and the NAEP national assessment test that measured the attainments of U.S. students.]

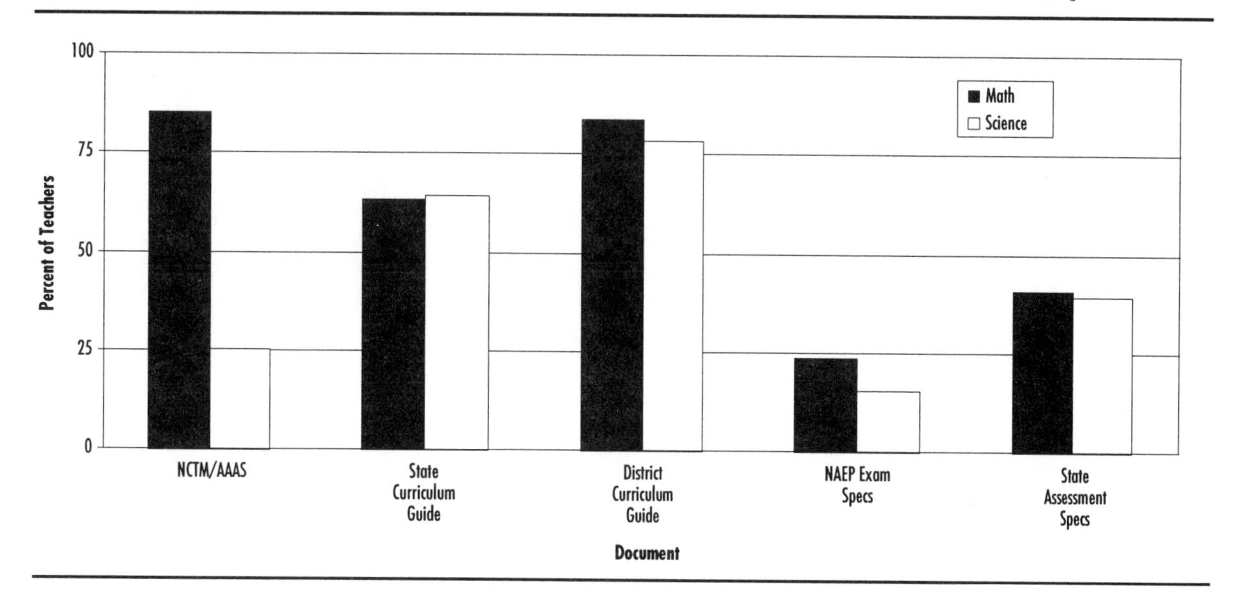

Exhibit 36.

Number of hours per week scheduled for mathematics and science teachers.

U.S. teachers are scheduled for far more hours per week than their counterparts in Germany and Japan. [This exhibit displays the number of in-class instructional hours for which mathematics and science teachers in the U.S., Germany, and Japan are scheduled. Items measured in periods were converted to hours using reported average length of a period.]

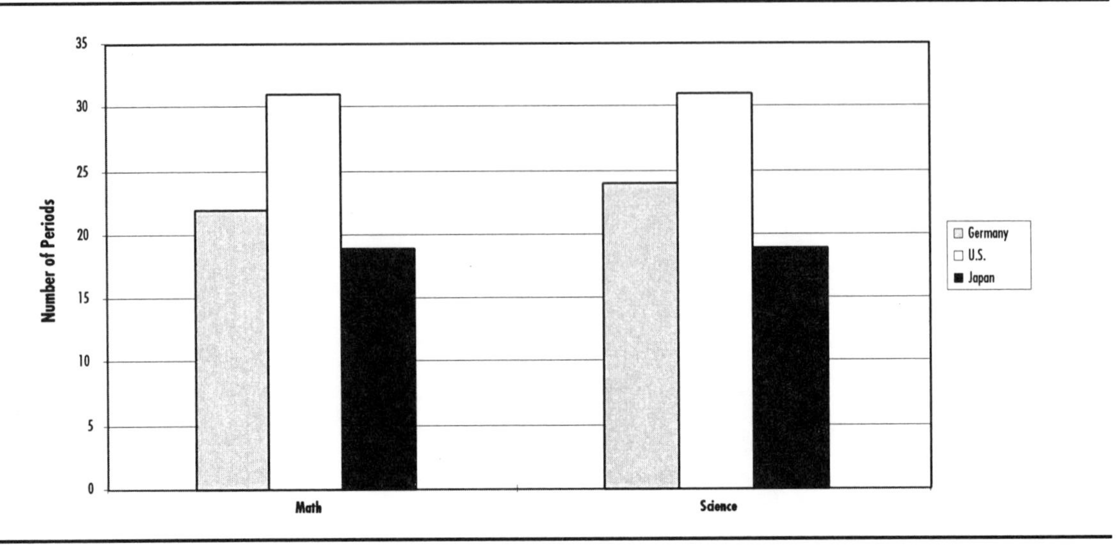

U.S. teachers. The German and Japanese periods per week appear to allow more time for a professional approach to instruction. This approach would include preparation, assessment of student work, additional individual work with students, consultation with colleagues, professional reading, and continued professional education.

Clearly the U.S. receives hard work from its teachers. They work long hours in difficult circumstances. However, if we wish to receive these teachers' best work, we must find ways to make their work more effective and create situations in which they can "work smarter" rather than merely "work longer."

Teaching: Summary and Concluding Remarks

U.S. mathematics and science teachers devote small amounts of instructional time to many topics. Even the few topics to which they devote more attention typically take up less than half of their instructional time. It seems likely that they make decisions based on pressing realities and frequent choices required by a demanding work environment. They perhaps settle for the first alternative that seems good enough to them rather than searching for the best.

Mathematics and science teachers instructional time allocations echo the inclusive, but breadth rather than depth, approach characteristic of our unfocused curricula and cautiously constructed, inclusive textbooks. Unfocused curricula and inclusive textbooks set few boundaries for instructional decisions. They make it easier for real teachers making real decisions in the real workplaces of U.S. schools to settle on alternatives and choices that are good enough, that is, to *satisfice*.[14] They often find these alternatives in materials such as textbooks. Teachers seem to consider textbooks, by their nature, to provide alternatives at least good enough for students, if not necessarily the best that they could find with more time.

In short, the realities of U.S. schools as workplaces and the limits on the professional lives of teachers seriously affect teachers' instructional decisions. These factors lead most teachers to echo the inclusive, exposure-oriented approach of U.S. mathematics and science textbooks. Unfocused, inclusive official curricula do not seriously challenge this approach. In mathematics, we must hope that the assumptions of distributed mastery are realistic. Mathematics achievements must be attained in an environment in which all major components determining instruction — official curriculum documents, textbooks, and teachers —make few choices that exclude either traditional or more recent reform-oriented contents.

Teachers at times choose paths more informed than the supporting and guiding materials provided them in curriculum guidelines and textbooks. It may be in this form, that recent mathematics reform recommendations have their greatest impact. At least some U.S. mathematics teachers indicate practices that reforms may have affected.

U.S. mathematics and science teachers are aware of and believe in more effective, complex teaching styles than they practice. They often have information that would help them do their work more effectively.

U.S. mathematics and science teachers work hard. There is little room for marginal improvements in effectiveness through longer or harder work. Such improvements would likely be at best quantitative rather than qualitative. The lengthy work week of typical U.S. science and mathematics teachers does not leave them much opportunity to pursue things that would qualitatively improve their teaching and move beyond survival-based satisficing.

We have yet to unleash the effectiveness of U.S. teachers. Effective teachers should not be unusual, nor should effectiveness require extraordinary efforts and dedication by teachers. Policies that make structural changes are much more likely to produce dramatic changes than merely providing information or seeking even more hours from teachers.

The proper question appears not to be, "Are our teachers doing their best?" They are not. They are, however, doing their best given the conditions in which they work. We should rather be asking, "Is our educational system doing the best it can for the children?" Are our curricula well designed to support instruction in powerful ways? Are our textbooks designed to support powerful teaching even if this requires daring changes?

We can, in fact, do better, but not by trying to extract "more of the same" from teachers while providing them "more of the same." Information- and motivation-based reforms and improvement policies alone will not bring fundamental improvements. This may set limits on the final impacts of the admirable recent reform efforts in mathematics and science education.

Any effective reform in this context will necessarily be systemic. The fundamental problem is not a conglomeration of individual problems. However, not every systemic reform will address the core of our problems simply because the reform is systemic. Appropriate reform must be structural and pursue focused and, meaningful goals. We may not be able to do everything and do any of it well. While a final conclusion awaits public discussion, it appears we must make choices, choices of which goals are more important and how many goals we can effectively pursue. We must find an intellectually coherent vision to guide our mathematics and science education.

Chapter 4

How Has Our Vision Become So Splintered?

We have seen clear evidence for a U.S. composite vision that is splintered in what it proposes should be done in mathematics and science education. This is seen in curricula without clear focus, in cautious and inclusive textbooks, and in teaching that echoes this uncertainty of goals and means. How has this happened? How have we become so disorganized in our views about what we must accomplish and how we must accomplish it? In part, our composite vision is inevitable, given that it is formed in thousands of sites and in a context of shared responsibility.[15] Our composite perceptions are further complicated by the fact that even each of these visions emanating from different sites is itself fragmented. We must try to understand each of these two sources of what complicates our insights and our plans for what we should do. Effective change requires us to examine hypotheses that explain how the current state of U.S. science and mathematics education has evolved.

Unlike some previous sections, what is offered here are hypotheses that seek to explain the data we have presented. Where our data support these hypotheses, we present these below. Are these suggestions *the* answers? Possibly not. However, they seem to meet the tests of being consistent with the data, sufficient (if not necessary) to explain the main findings of our investigation, and, hopefully, insightful enough to further discussion and the search for explanations of the data we have presented.

Who Is In Charge Here?

We will not escape the fact that our educational visions are shaped in a context of distributed and shared responsibility. We are educational federalists.[16] Educational federalism has traditionally been a loose union with more responsibilities and rights reserved to states and local districts than is true for many other areas of the common good.[17]

The result is a federalism of guiding visions that typifies U.S. educational practice both past and present.[18] There are no single coherent, intellectually profound, and systemically powerful visions guiding U.S. mathematics and science education. An ephemeral aggregate of fragmentary intentions stands in coherent vision's place in the U.S. as it does in few other nations.

Reform efforts in mathematics and science education have offered coherent and powerful visions. However, they have yet to achieve the consensus and lack of variant interpretations that would allow them to be single visions, one for mathematics and one for science, that consistently and profoundly shape U.S. educational practice. Perhaps we do not need consistency in guiding visions. Perhaps the values of diversity outweigh those of focused, unified pursuit of a few key goals. However, while some would have this remain an empirical question, given the effects of unclear visions and goals revealed by our data, the importance of seeking more coherence should certainly be a question for public discussion.

We have no shortage of visions of how U.S. mathematics and science education are "supposed to be." Partly this is because we have no shortage of sites articulating visions. In the U.S., shareholders in the "official" vision enterprise include more than 15,000 local school districts and boards, 50 state education agencies, and various federal offices, committees, boards, and administrators.[19] Others with a stake in the enterprise join official shareholders. These include textbook and test producers, members of professional organizations in mathematics and the sciences, teachers' organizations, special interest groups with educational goals, governmental officials at all levels who state policy broadly but not in detail, etc.

In the face of this federalism of visions, does it even make sense to speak of a U.S. educational system, a U.S. vision for mathematics and science education, or national intentions in these areas? If we mean by a U.S. "system," a voice or set of intentions that come from a single, official source, then there is no U.S. educational system, no one voice, no single set of intentions. There are only educational systems, many voices, and multiple sets of intentions across the U.S. However, some individual U.S. states are as large as some TIMSS countries and educational practice there is shaped by goal-driven agencies just as are national education ministries in other countries. While states have considerable independence in the U.S. system, they are not isolated, and educational counterparts communicate regularly with each other. There are some bases for underlying commonalities that are not obvious when we consider only the "degrees of freedom" in U.S. mathematics and science education.

Does it make sense to use only this narrow sense in considering a national education system and vision? Such concern seems limited to education. The U.S. does not directly compete as a single body in international marketplaces. Individual U.S. corporations, products, and service providers compete with individual corporations, products, and service providers from other nations. Yet we do not hesitate in speaking of whether the U.S. is "competitive." The sources for economic policies and strategies are many and varied, but we speak as if they made a meaningful national aggregate. That is, we speak as if there was a composite "yield" from these diverse efforts that we can assess and seek to improve.

We are willing to speak of national aggregates and yields in other areas. Why do we hesitate to do so in education? We do have a composite yield for U.S. educational efforts. We often assess this yield by measuring student achievements for national samples — for example, in the continuing work of the National Assessment of Educational Progress (NAEP). We assess this same composite yield — the aggregate output of U.S. educational systems' efforts at all levels — in cross-national studies such as the International Assessment of Educational Progress (IAEP), the Second International Mathematics Study (SIMS), the Second International Science Study (SISS), and, currently, the Third International Mathematics and Science Study (TIMSS).

Educational achievements are outcomes of educational processes. If we measure composite educational achievements, we must regard these as the "yield" of some process or effort. We obviously must hesitate to speak simply of *a* national educational system or vision given our distributed sharing of responsibility for providing schooling. However, we must at least be willing to speak of a *composite* U.S. educational system and vision. We must aggregate carefully the visions and goals of individual participants to find a *de facto* composite vision.[20] Without these composite "inputs," measuring yield as a composite "output" is enfeebled and an inappropriate double standard is maintained. It is exactly these composite inputs that we characterized by the data we presented earlier.

We have avoided certain terms — 'curriculum,' 'goals,' and 'standards.' National curricula, national goals, and national standards are the subject of continuing heated public debate. Many consider them to imply official policies, mandatory guidelines, and decisions made at the national level. We wish to distinguish carefully between *composites* of nationwide actions and *official actions* at the U.S. national level. We speak only of composites for the U.S. in discussing our findings.

The recent history of national goals and standards has been both accompanied by heated public debate and filled with frequent reversals of direction. Just over five years ago, at the instigation of state governors, major national educational goals were articulated.[21] Even at this early stage, official statements spoke of the "nation's education system" but explicitly recognized educational federalism.[22] As part of achieving these goals, national panels were created and links to assessment sought.[23] Relevant professional organizations' activities paralleled (and, in some cases, preceded) these official national actions.[24] A recent study of activities in the states shows that these national and professional organization actions have affected the states. The studies show that the states are working actively to revise their mathematics and science curriculum documents at the state level, and to articulate "standards" of desired performance.[25]

Even allowing for care not to suggest a national curriculum, these concerns for national goals and national standards proved controversial.[26] The public debate on goals and standards, national or otherwise, continues to date.[27] Clearly, we do not suffer from lack of visions for how mathematics and science education are "supposed to be" — although we may suffer from a lack of vision. One author described the situation by writing, "The sources of... standards seem as plentiful as stars in an inky sky."[28] Any serious reform proposal in today's situation risks being just one more "voice in a babel"[29] confronting state and local educational agencies. This, in part, lies behind the continued struggle to secure shared efforts orchestrated by discussion at the national level,[30] efforts that remain a source of controversy.[31]

Organizational Issues

Unfortunately, these many visions from their many sources often at least partly conflict. This contributes in mathematics and science education to inclusive but unfocused curricula, textbooks, and teachers' efforts. These curricula, books, and teachers are undoubtedly driven by goals, but not effectively so. How should we understand our distributed responsibility for school science and mathematics. In today's climate emphasizing decentralized authority for educational matters, we must surely consider the federalism of guiding visions for mathematics and science curricula as "loosely coupled."[32]

Formally, this situation is a loose coupling of several relatively independent "actors." What is the basis for their actions and decisions and how does this affect the aggregate coherence of emerging policies? One possible organizational model is a classical, "rational actor" model. This model considers each actor to behave rationally in making individual decisions with an eye to the cumulative effect of those decisions on the whole.[33] The strong mutual concern and sense of shared responsibility implicit in this model does not seem characteristic of our situation today.

An alternative model seems likely to be more germane. This is an "organizational process model"[34] that views government as a conglomerate of many loosely allied subunits each with a substantial life of its own. This certainly seems more characteristic of the loose federalism guiding science and mathematics education locally, at the state level, and nationally. This is especially true when we include secondary actors, such as professional organizations and textbook and testing organizations, in the picture. Each "actor" pursues his or her own "life" — his or her goals, visions, plans, processes, and efforts to satisfy those to whom he or she is accountable. The aggregate effect of these separate lives is a secondary concern for most.

The decisions, policies, and documents that flow from this conglomerate of "subunits" should be considered not so much as deliberate choices contributing to an aggregate effort. Rather, we may better consider them "outputs of large organizations functioning according to standard patterns of behavior."[35] Thus, the parts of our federalism not only act with primary reference to their own internal life, but they operate by traditional patterns. These patterns vary in how integrally they include attempts at "rational" decision-making and concerns for aggregate effects beyond their own particular concerns.

In the U.S., states and school districts set up curriculum frameworks, standards, objectives, and other curriculum documents. Commercial publishers develop textbooks for use in science and mathematics classrooms. National associations suggest reforms. Test publishers develop tests that have curricular implications. Teachers make day-to-day classroom plans and implement them. The federal government develops a "national report card" based on NAEP findings that has implications for the success of curriculum efforts. Some national boards and programs struggle to influence coherence in the aggregate vision of science and mathematics education. However, little of this is coordinated, and what coordination exists is deliberately not mandatory.

In this kind of situation — various organizations loosely joined into a larger enterprise (U.S. science and mathematics education and "yield") — the organizational process model predicts that the component organizations will not always work towards common goals. They will not always aim at producing important combined results. Individual organizations tend to act parochially and are rightly concerned for the agendas of their own responsibilities. The actual results are a consequence of "innumerable and often conflicting smaller actions by individuals at various levels of bureaucratic organizations in the services of a variety of only partially compatible conceptions of national goals, organizational goals and political objectives."[36]

Others[37] speak of problems so complex that they are split or factored into parts that are then parceled out to various organizations. This seems more characteristic of an aggregate rational actor model than of the organizational process model most directly applicable to our federalism. No central directing agency factoring the complex problem of U.S. mathematics and science education policy into pieces tackled separately. The system distributes responsibilities, not modular tasks to be integrated into some larger whole. Instead, the U.S. has a conglomerate of independent, loosely interrelated "actors" each pursuing their own solutions. The aggregate result is a set of disparate solutions that partly coincide, partly vary. This seems clearly to reflect one source of the splintering we find in the guiding visions of U.S. science and mathematics education.

Many people have a tendency to assume a model of rationally acting organizations that somehow share a coherent enterprise. This would allow clear assessment of cause and effect and the ability to assign praise and blame to various "actors" for why things are as they are. Many in the U.S., for example, would seek to blame teachers or textbook publishers for the current state of mathematics and science education. However, the more realistic model seems to be one of independent organizations pursuing their own properly parochial goals, but often with little concern for or insight into the composite U.S. science and mathematics curricula.

Formal mechanisms of coordination — either by regulation or rewards for selected behaviors —have proven politically sensitive and are in limited use. Even were such coordination to exist, there is no coherent intellectual vision of U.S. mathematics and science education to inform it. The broadly systemic problem, with its structural barriers to progress, has made the pursuit of intellectually coherent solutions so unrewarding that none currently exist, and our guiding visions are fundamentally splintered. Even when a reasonably coherent vision emerges as a voice for reform—for example, the NCTM *Standards*, the AAAS *Benchmarks*, and the

National Academy of Science's National Research Council's *Science Education Standards* — in our organizational context it becomes simply one more voice in a "babel" of competing voices. It becomes so much so that it is hard for individual official actors to separate "signal" from "noise" or to prioritize the voices to which they will attend. In such a systemic context, splintered visions are likely to remain.

Our point is not to condemn or bemoan our tradition of shared responsibility for common education. Rather, we wished to characterize carefully the context in which we must seek and implement any needed changes. The weight of public discussion may be that data such as those we have presented do not call for action. However, if action seems called for, it must be appropriate to our educational federalism. If discussion suggests the need for more powerful and coherent guiding visions, they must be sought in processes that will achieve wide consensus necessary for change in our context. Any serious attempt at change in U.S. science and mathematics education must be deeply structural. Our point has been to clarify how complex such structural change will be. Were such change simply attainable, it is unlikely that U.S. curricula, textbooks, and teaching would have come to have their present unfocused, inclusive character.

Incremental Assembly: Mass Production, Mass Education

Our point thus far is that the organizations in our system of distributed responsibilities for science and mathematics education act according to their own visions and agendas. Given this comparative independence, why do these visions replicate content fragmentation so often and consistently that it is characteristic of U.S. *composite* intentions? Certainly one or more other sources of splintering must affect each site of individual planning to produce the widespread fragmentation our data revealed. What insights can we gain into other sources for our splintered visions?

The U.S. is an industrialized society, a product of the Industrial Revolution and its particularly American form in mass production on efficiently run assembly lines. When we consider some of our greatest national successes — particularly our economic successes — we are all children of Eli Whitney and Henry Ford. This may change in a service-oriented "information age," but the assembly line and mass production are still integral parts of our national heritage. While initially conceived in producing physical products, many of these ideas were adapted to producing intellectual products as well. These notions are subtly embedded in our thoughts about solving any large-scale problem. For example, in describing the Ford Foundation's efforts to improve adult education, one author wrote, "Socrates had been moved from the Academy in Athens, to a Detroit-like assembly line..."[38] What was true about the intellectual impact of mass production on adult education is also true more generally of the impact of mass production on mass education.

To gain further insight into how mass production has given us guiding images that have shaped our approaches to the problems of mass education, we need to consider briefly some of the "principles" underlying assembly line production. We will call such assumptions or "principles" the tenets of a model *of incremental assembly*. Four of these principles seem particularly relevant to the repeated fragmentation characterizing the splintered vision of today's U.S. science and mathematics education.

First, as Eli Whitney discovered in designing more powerful production of muskets under government contract (probably the first American example of effective assembly line production),[39] a manufactured item is divisible into individual parts. This allowed one worker to focus on making one particular interchangeable part quickly and well. This contrasts with making all parts and assembling a finished item as a traditional craftsman would have done. The principle for incremental assembly is that

- Any complex object, process, or activity can be partitioned into component parts, subprocesses, or subactivities.

Applied to education, this principle suggests that complex skills, ideas, and bodies of knowledge can be broken into simpler skills, concepts, and facts that we may approach independently.

Second, to achieve efficiency by having more than one person work on the same product or project (as also demonstrated by Whitney), we assume that

- All parts, subprocesses, or subactivities of the same type can be produced uniformly and interchangeably.

This allows the benefits of worker specialization and of having more than one worker for a complex task. Uniformity and interchangeability seem essential for education in a highly mobile society in which the same child may study grade 3 mathematics in one city and grade 8 mathematics in another.

Third, to produce a complete object, process or activity (as Whitney proved to skeptics), we assume that

- All required parts, subprocesses, or subactivities when completed can be assembled and integrated to form the complete, complex object, process, or activity.

Without this assumption, there would be no point in partitioning complex objects or processes into simpler, more limited pieces. The goal is always the final object or process. The parts are the means.

Fourth, to allow incremental assembly, we must also assume that

- Different required parts, subprocesses or subactivities can be completed by different workers, either simultaneously or at different times, to be available when needed.

This allows the completion of a complex production in stages, rather in one unbroken sequence of actions.

These principles of incremental assembly seem to underlie many and varied aspects of American life — industrial production, modularized large-scale computer software development, psychological behaviorism, etc. We suggest that at least some of these principles also underlie U.S. school education activities, in particular those aimed at mathematics and science learning.

The principles of incremental assembly in many forms have been implicitly a part of American education for decades. For instance, one form was the emphasis on behavioral objectives in the 1960s. Behavioral objectives, with an implied link to behaviorism and stimulus-response psychology, seemed on the surface a simple notion: Define what is to be accomplished in learning so precisely that at any point a specific objective guides instructional activities. As a result, the accomplishment of that objective can be measured with fair objectivity since the goal has been expressed in terms of specific behaviors.

That simple notion had a number of hidden perils, however. First, specifying precisely all that students were expected to do "required the construction of *hundreds* of specific objectives"[40] (italics original). The evidence of applying these ideas to mathematics instruction, at least in the form of programmed instruction,[41] clearly involved breaking skills to be mastered into simpler subskills. When all necessary subskills were mastered, they were to be integrated into the more complex overall skill. Programmed instruction also held that students should be asked only simple questions that they had a high probability of answering correctly. If materials were tried and questions or tasks were found to have "unsatisfactory" error rates, those tasks were broken down further into even simpler questions or tasks.

Instruction today does not proceed in such a doctrinaire, methodical fashion. However, the legacy of these methods — and the assumption that incremental assembly is a useful approach for mass education — remains with us.[42] We still find extensive lists of very specific objectives (no longer in a doctrinaire behavioral form) in official curriculum guides today. Further, behind these objectives are the assumptions of incremental assembly, the assumptions that complex wholes can be assembled from incrementally accumulated parts. These assumptions free curriculum planners, even encourage them, to fragment content. This approach seems particularly suited, even necessary, to an American society that is far more mobile than those in most other countries.

We commonly assume that we can break "big ideas" into "little ideas" and complex learning into simpler learning tasks. We assume that we can disassemble ideas, skills, and processes into essential, but simpler pieces. We assume that instruction can produce these pieces uniformly and interchangeably. That is, we assume that if a student masters a simpler idea, skill, or process, it does not matter from whom, at what time, or by what means he or she attained that mastery. The mastered simpler learning is available for more complex learning. We assume that these "pieces" can be developed sequentially, by different instructors, in different grades, from different textbooks, etc. We assume that, once mastered, we can assemble them into the more complex ideas, skills, and processes that are our true goals. Without such assumptions, distributed mastery as seen in our data would be untenable. Even the spiral curriculum approach set out by Bruner[43] involves a more sophisticated version of incremental assembly.

Reflections of Incremental Assembly Assumptions

Do versions of incremental assembly underlie U.S. science and mathematics education today? This is hard to establish directly since such assumptions are usually implicit. However, incremental assembly is reflected in certain characteristics of curricula, textbooks and teaching. If we can identify some of those reflections we can search for their empirical presence in mathematics and science education practice. They will form supporting evidence that incremental assembly is an implicit assumption — and, given its nature, a further, ubiquitous source of fragmentation and splintering.

Curricular reflections: Unfocused curricula. If incremental assembly assumptions underlie the thinking that shapes a curriculum guide, they should lead to (1) a large number of topics meant to receive limited curricular attention (the smaller pieces to be assembled), (2) simpler expectations for most topics and small amounts of time intended for their study (since the pieces are small), and (3) extended durations as topics are developed over several years by assembling pieces and by elaborating originally learned simpler versions. Our earlier discussion of the U.S. composite science and mathematics curricula and of official state curriculum guides found these characteristics broadly present in U.S. science and mathematics curriculum guides. It was exactly these characteristics that caused us to identify the U.S. curricular intentions analyzed as "unfocused." However, this results at least in part because the standardization necessary for incremental assembly is not achieved in U.S. mathematics and science education.

We do not claim we know the precise factors influencing the thinking of the many official participants whose work produces these characteristics again and again. In mathematics, it may be ideas of distributed mastery or spiraling to gradually produce deeper, more elaborate versions of the things learned. In science, it may be spiraling in curriculum, progressive refinement and elaboration of central ideas and processes, or some other motive. What seems undeniable is that curriculum after curriculum in science and mathematics freely fragments content. They break major topics into large numbers of more specific topics. This requires that many topics be covered each year, that little time typically be planned for most topics, and what is to be done with most topics is restricted to the less demanding. Only occasionally do curricula include more demanding, integrated expectations.

We also do not claim that this fragmentation occurs only in the U.S. However, the curricular data make it clear that the U.S. generally does more of this, does it more consistently, and does it more extensively. We find a level of fragmentation in our unfocused curricula not often matched. Our curricular vision is certainly splintered in this way and, usually, far more so than most other countries.

Textbook reflections: Fragmented content, limited expectations, and short attention spans. If textbooks reflect these underlying assumptions of incremental assembly, a textbook series would be free to gradually develop larger pieces of content from smaller pieces. A series would also be free to distribute this development in the books for several grades. This may be done as a cautious response to unfocused curricula and conflicting demands that put a premium on inclusive approaches to content. For whatever reason, it *is* done. A characteristic problem for textbook design would be how to efficiently handle a large number of fragmented topics so that each book in a series provides some useful pieces in each topic area. The design might provide for a variety of performance expectations but would use a large proportion of the less demanding expectations. A textbook would need to be designed to support instruction effectively (e.g., to include a large number of exercises for each of the fragmented, multiple topics). Because of reform concerns and to be acceptable in a wider share of the market, a textbook would need to be inclusive. It would have to include both traditional and newer contents, and to include at least a "spice" of more demanding tasks and activities for students.

We have seen earlier that U.S. science and mathematics textbooks are very inclusive. They include many topics, both traditional and reformed. They allocate limited space to most topics. They provide for varied performance expectations, but with a strong core emphasis on more routine, less demanding performances. What we have not seen is what this fragmented approach does to the detailed *organization* of textbooks. They face the need to balance supporting an inclusive, fragmented curriculum with presenting material in an interesting fashion that holds student attention. How do mathematics and science textbooks achieve this?

We gain some insight into this by examining the sequence of contents in successive blocks (small analytical segments) of textbooks. By comparing sequences for textbooks from different countries, we can see whether U.S. science or mathematics textbooks are organized differently. A map of the content sequence of a typical Japanese Population 2 mathematics textbook (see Exhibit 37) shows that it focuses on a comparatively small number of mathematics topics. There are long, contiguous sequences of blocks for the same content for several topics. The Japanese text focuses mostly on a few topics and provides extended attention to them. There is some, but comparatively little, skipping from topic to topic in successive blocks.

The picture of a typical U.S. grade 8 mathematics textbook is quite different (see Exhibit 38). This textbook provides material on a comparatively far greater number of mathematics topics than the Japanese book. There are some sequences indicating somewhat more extended attention to a topic. However, compared to the Japanese text, there are fewer of these sequences (fewer topics focused on). The U.S. textbook's sequences are shorter (less connected attention given to emphasized topics). The U.S. textbook's sequences have more breaks (that is, it attends

Exhibit 37.

Schematic overview of a Japanese Population 2 mathematics textbook.

This Japanese textbook is characterized by use of only a limited number of topics and by long dense row segments indicating “blocked” coverage focused mainly on a single topic. [Our goal in this exhibit is a general impression, not complex detail. We mark every block (small segment identified as a unit of analysis). Each column corresponds to one block. We present the blocks sequentially through the pages of the textbook. Each row represents a different content topic from the TIMSS’ mathematics framework. Shadings in a column show the topics identified for that block. Numbers correspond to codes from the mathematics framework - see reference.]

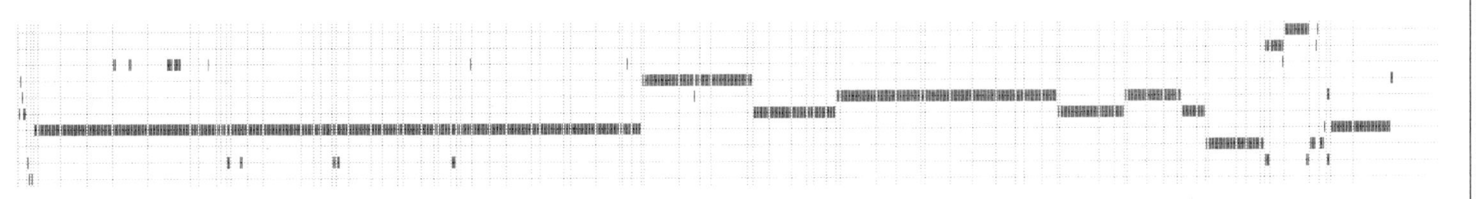

Exhibit 38.

Schematic overview of a selection from a U.S. Population 2 mathematics textbook.

The U.S. textbook covers far more topics than the Japanese text (as shown by the numbers of rows for each). There are fewer row segments that indicate more extended attention to a topic. There is more skipping among topics. [Our goal in this exhibit is a general impression, not complex detail. We mark every block (small segment identified as a unit of analysis). Each column corresponds to one block. We present the blocks sequentially through the pages of the textbook. The four diagrams represent one book partitioned into four arbitrary pieces going from left to right and top to bottom. Each row represents a different content topic from the TIMSS' mathematics framework. Shadings in a column show the topics identified for that block. Numbers correspond to codes from the mathematics framework - see reference.]

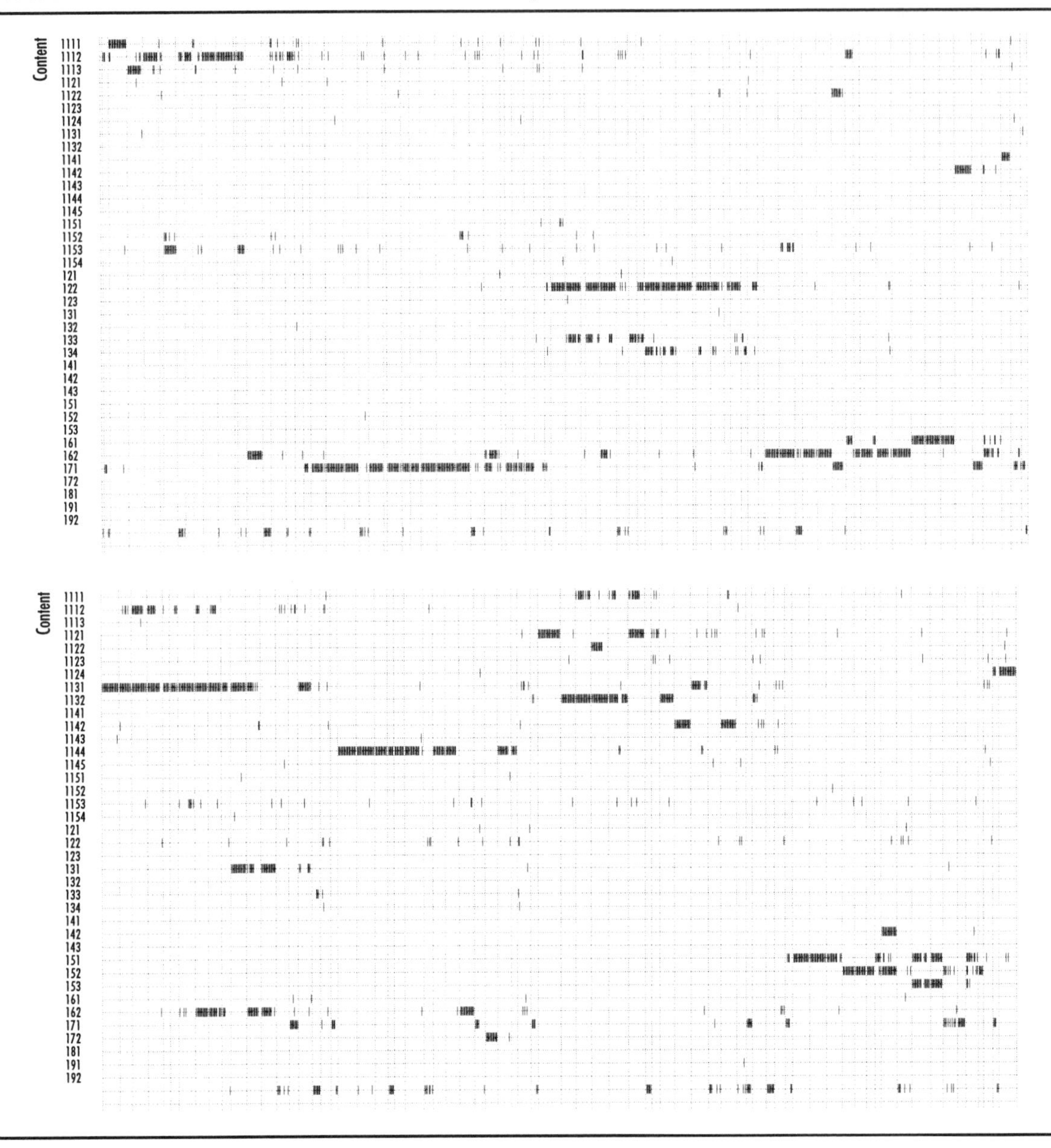

Exhibit 38. (cont'd)

Schematic overview of a selection from a U.S. Population 2 mathematics textbook.

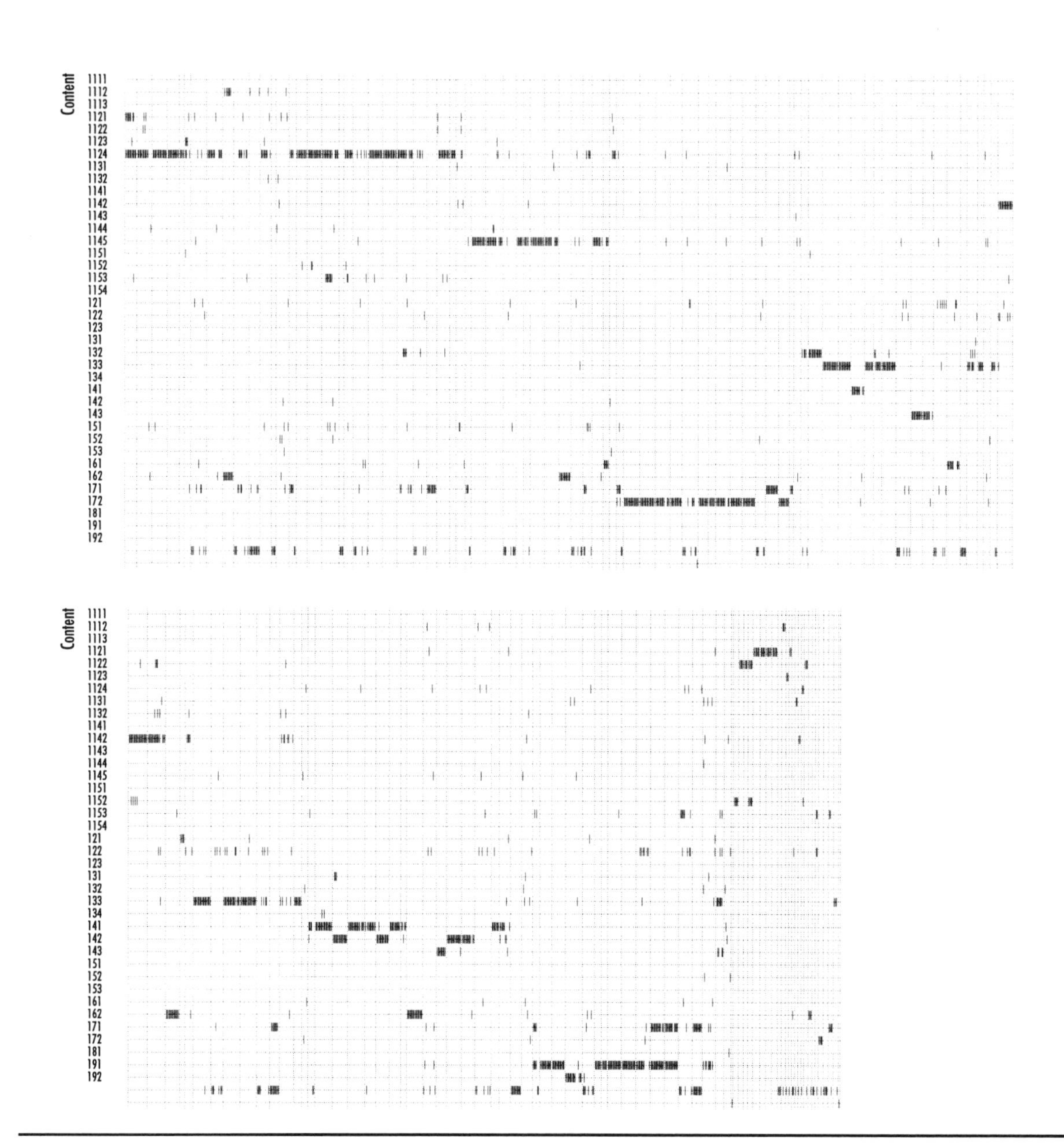

to other topics even in a sequence primarily devoted to one topic). The U.S. book is much less focused than the Japanese book in lesson-by-lesson organization. There is far more skipping among topics in successive blocks.

These are fairly typical data for textbooks (although this analysis is on-going). U.S. books seem to have been more inclusive and less focused within each book, not just across longer curricular sequences. U.S. textbooks clearly use variety and "spiraling" presumably to help hold student attention. This is done at the price of focusing, of treating content in longer sequences. For whatever reasons, this U.S. textbook was fragmented even at this "micro-organizational" level. Clearly this makes it more demanding for students to link contents together.

The percentages of different performance expectations used by grade 8 textbooks also reveal comparative differences. (See Exhibit 39 and Exhibit 40 for performance expectations for one representative topic each from mathematics and science.) Almost every country, for both mathematics and science, emphasizes simpler expectations such as 'knowing,' 'understanding,' and 'using routine procedures.' More complex performances of investigating, reasoning, 'communicating,' etc., are only rarely used. The data are similar for other topics.

Why is there such consistent emphasis on less demanding performance expectations in mathematics textbooks? While the reasons are unclear for other countries, the emerging pattern for U.S. textbooks suggests one explanation. If topics are presented in smaller pieces, perhaps over several grades, and, especially, as a part of incremental assembly, it is important that these small topics be mastered with sufficient uniformity that they can be counted on for later "assembling" into more complex learning.

Only when a topic is considered as a focus or in a more holistic, extensive, or intensive way is it reasonable to include more demanding performance expectations for it. Only then are we likely to go beyond simple factual knowledge and routine procedures. Fragmented topics reasonably seem to be coupled with simpler learning demands and expectations. Only when — and if — topics are finally assembled into more complex, integrated ideas and skills would more complicated learning demands become appropriate. A side effect of a steady diet of fragmentation is that more complex demands seem often to be indefinitely delayed.

Instructional reflections: How teachers organize lessons. U.S. grade 8 teachers' use of activities within an instructional period seems also to reflect the effects of fragmentation and its concomitants. U.S. teachers use far more activities in a single lesson than do Japanese and German teachers. Around 75 percent of the grade 8 mathematics teachers in all three countries reported using four or more activities in a lesson. However, more than 60 percent of U.S. Population 2 teachers reported using six or more activities. In Germany only 25 percent of the grade 8 teachers reported using six or more activities, and even fewer reported doing so in Japan. The typical activity pattern of a U.S. lesson seems to echo the variety and skipping typical of U.S. textbook organization. This is an instructional approach more appropriate for fragmented content than for more focused or integrated content. This is true for both mathematics and science, but less so for science (see Exhibit 41, Exhibit 42, and Exhibit 43).

Exhibit 39.

Grade 8 mathematics textbook performance expectations for 'Congruence and Similarity.'

All textbooks tend to use the least demanding expectation the most, except for Germany's use of 'investigating and problem solving' and Japan's use of 'mathematical reasoning.' U.S. textbooks make some use (more than the international average) of complex procedures and communication. [The columns represent the average percentage of grade 8 mathematics textbook blocks indicating 'congruence and similarity' that also indicate the column's performance expectation. We give data for the international average, the U.S., Germany, and Japan.]

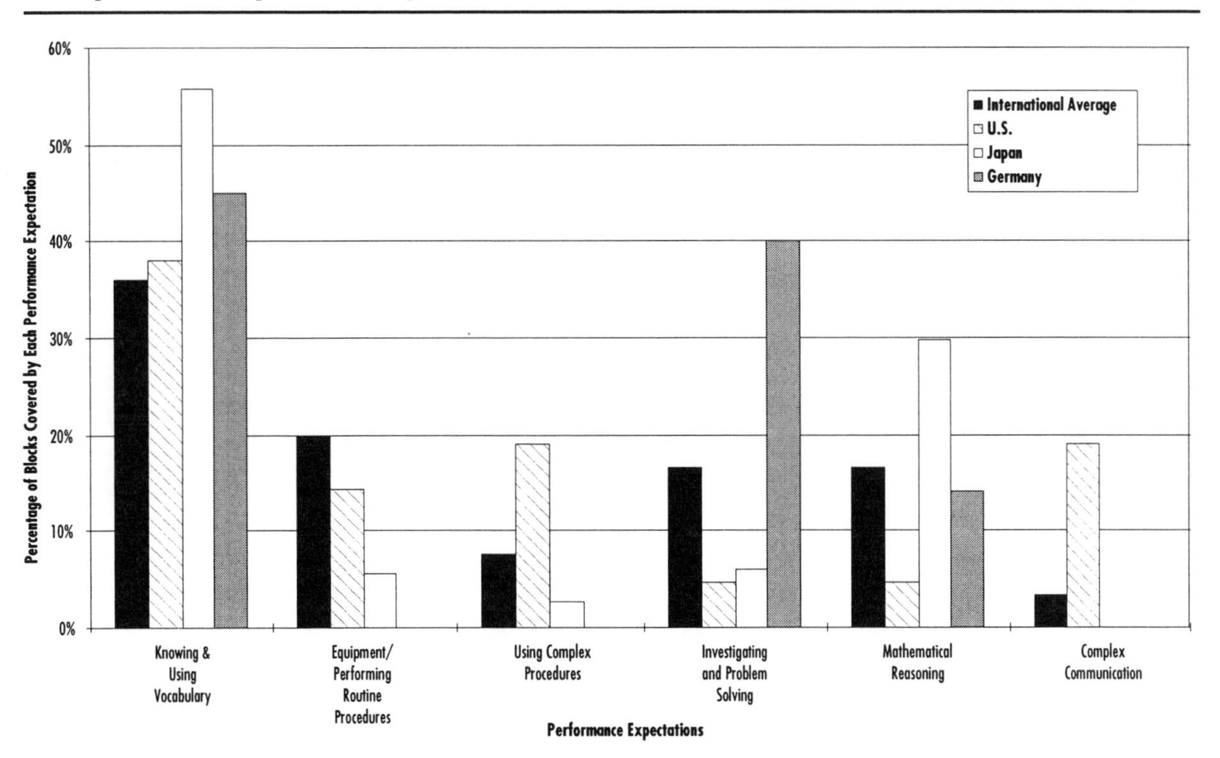

* Missing bars indicate that the performance expectation was not included for this topic.

Exhibit 40.

Grade 8 science textbook performance expectations for 'Chemical Changes.'

All textbooks tend to use the least demanding expectation almost exclusively, except that Japan used only 'understanding complex information' rather than 'understanding simple information.' Germany made more use of routine procedures. In all cases the emphasis was almost entirely on understanding. [The columns represent the average percentage of grade 8 science textbook blocks indicating 'chemical changes' that also indicate the column's performance expectation. We give data for the international average, the U.S., Germany, and Japan.]

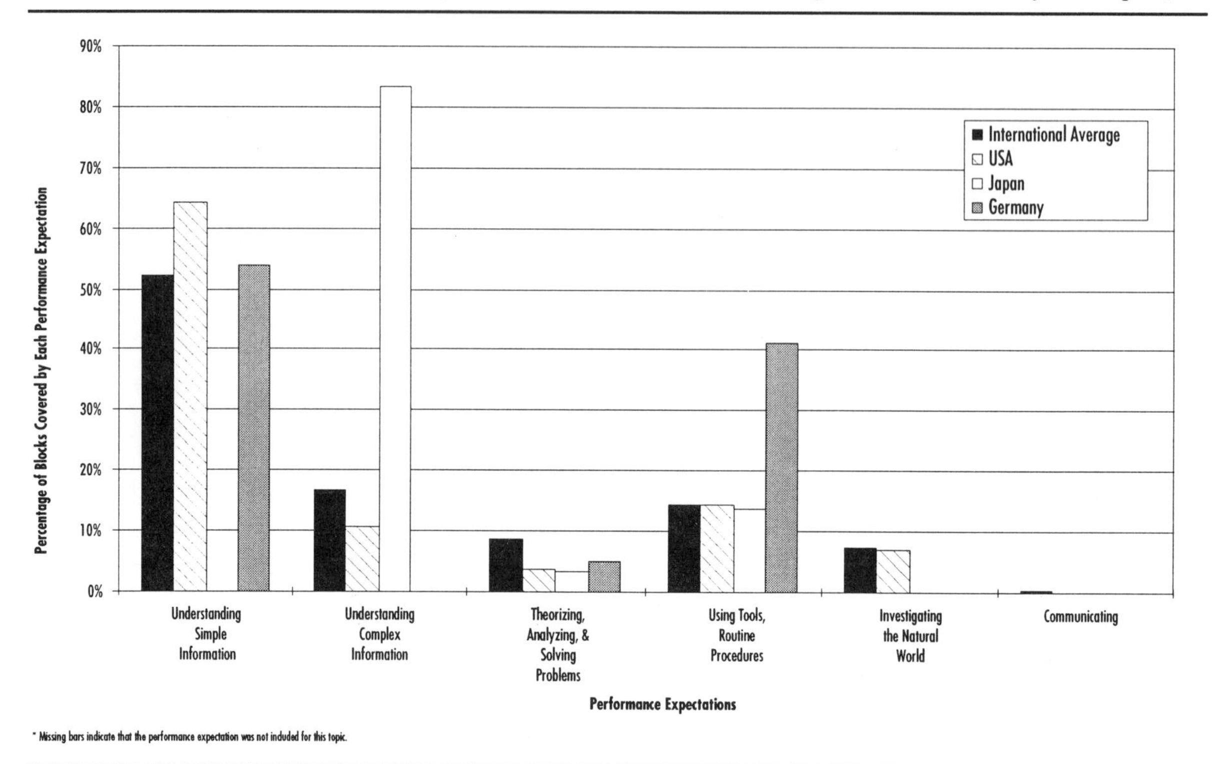

* Missing bars indicate that the performance expectation was not included for this topic.

Exhibit 41.

Number of activities used by grade 8 mathematics teachers in one period.

Most teachers reported using several different activities within an instructional period. However, U.S. teachers appear typically to have used far more than did teachers from Japan or Germany. [Grade 8 mathematics teachers in the U.S., Germany, and Japan reported how many activities they typically used in one mathematics lesson. The columns are for different numbers of activities. The height of each column indicates the percentage of grade 8 mathematics teachers that reported using the number of activities for that column.]

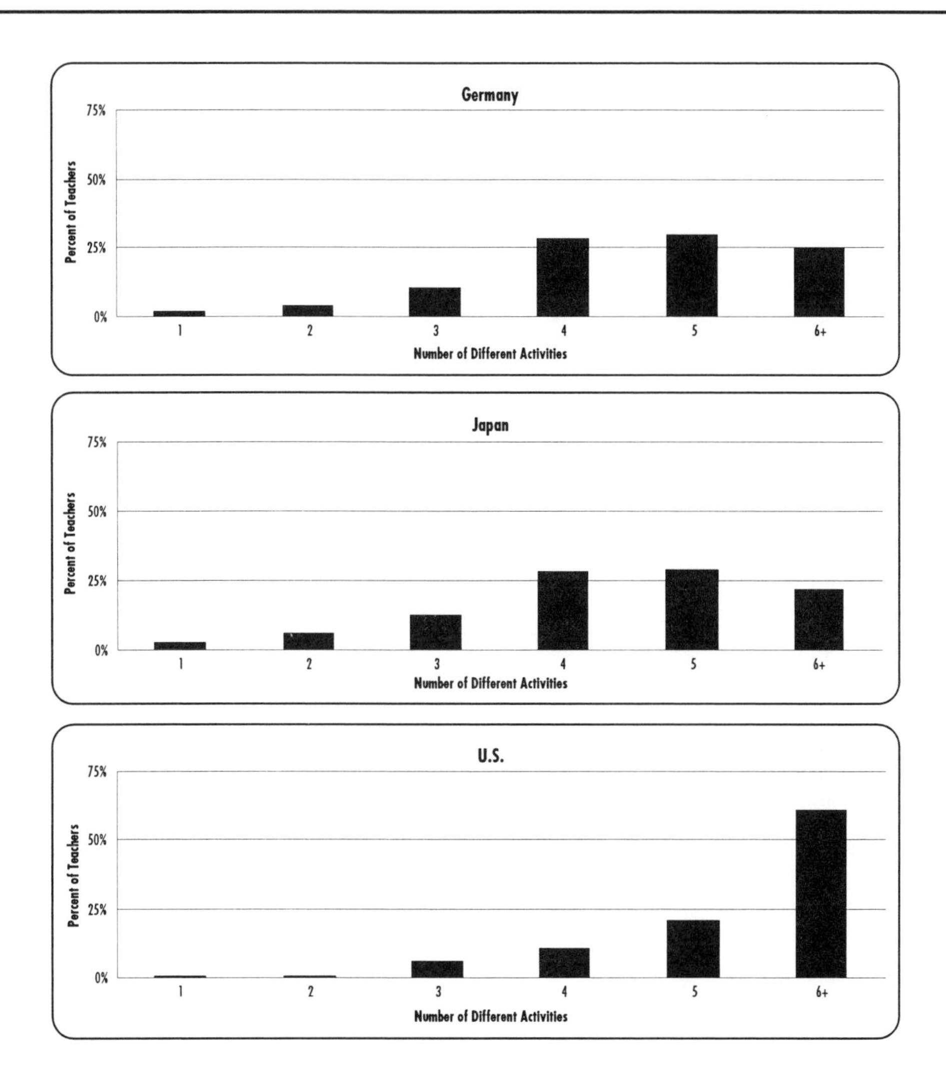

Exhibit 42.

Cumulative distribution of the number of activities used by grade 8 mathematics teachers in one period. US grade 8 mathematics teachers report consistently using more different activities than their counterparts in Germany and Japan. [We represent each country by a curve showing the cumulative proportion of their teachers using different numbers of activities. For any point on one of the curves, the point's distance to the right indicates how many different activities are being considered. The point's height indicates the percentage of that countries' grade 8 mathematics teachers who use that many or fewer activities in a typical lesson. Curves that lie further to the right indicate typical use of more different activities.]

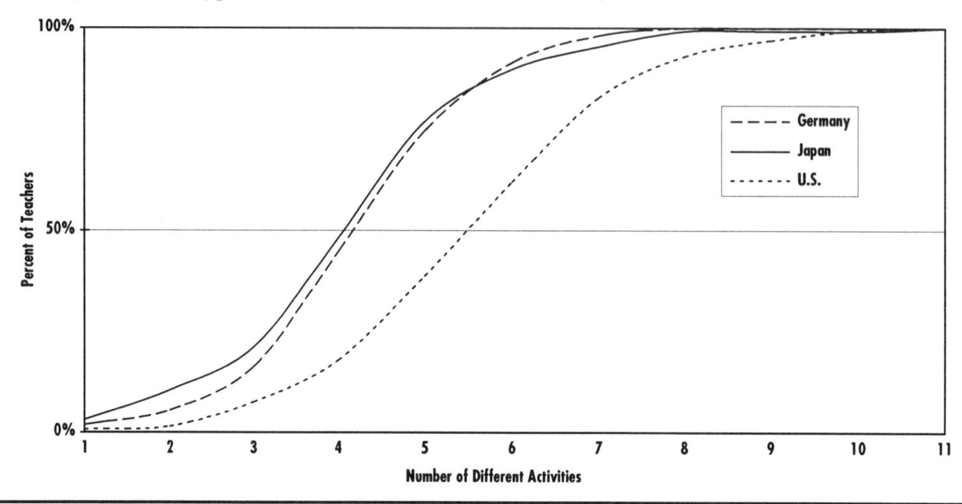

Exhibit 43.

Cumulative distribution of the number of activities used by grade 8 science teachers in one period. U.S. grade 8 science teachers report consistently using more different activities than their counterparts in Germany and Japan, but the differences are smaller compared to those among grade 8 mathematics teachers. [We represent each country by a curve showing the cumulative proportion of their teachers using different numbers of activities. For any point on one of the curves, the point's distance to the right indicates how many different activities are being considered. The point's height indicates the percentage of that countries grade 8 science teachers who use that many or fewer activities in a typical lesson. Curves that lie further to the right indicate typical use of more different activities.]

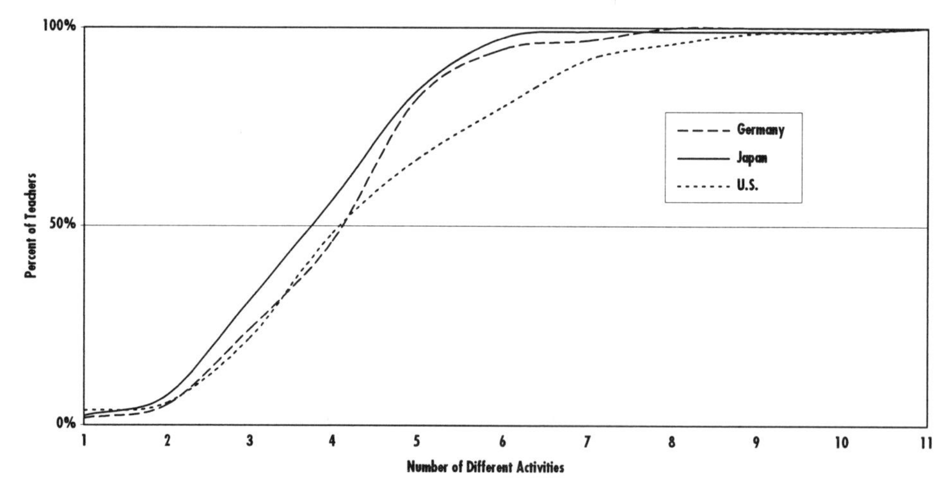

Observational data gathered in six countries as a part of TIMSS instrument development[44] provide qualitative information supporting these more quantitative data. U.S. teachers not only break content into small fragments (typical of their curricula and textbooks) but also break instructional periods into many small activities. We observed them moving frequently from one activity to another, more so than in the other five countries. Certainly this seems part of an effort to keep U.S. students engaged with the content studied. However, something of a vicious circle seems involved here. These efforts to secure engagement are made necessary by an approach that fragments content to the point that any one piece has little inherent interest. Teachers must supplement inherent interest by efforts to secure engagement and build interest.

Will recent reform recommendations remedy this pattern of frequent transition among activities? There have been both perennial and recently renewed concerns to adjust to individual student differences. In recent reforms, the suggested approach has been a constructivist approach that puts a premium on building from what students already know or believe that is relevant to a subject. Inherent in such a strategy is the need to determine what students bring to a lesson and then adjust to that in the instructional activities that follow. This requires several sorts of activities. However, if genuine contact is made with student experiences and interests, the need to artificially stimulate engagement with less-than-engaging content would be lessened. However, constructivism offers its own kinds of fragmentation by basing content integration not on design but on having desired goals emerge from opportunistic and wise use of student beliefs and interests.

The observational data also suggest that U.S. teachers rely on seatwork more heavily than did teachers from the other five observed countries. Seatwork is individual work by students at their own seats during class time. U.S. teachers also relied more heavily on "homework" begun in in-class time. Only Norway was similar to the U.S. in this regard. Quantitative data from TIMSS support these observational findings (see, for example, Exhibit 44). Grade 8 U.S. teachers use both seatwork (in-class exercises) and in-class homework. Japan and Germany use even more time for in-class exercises than does the U.S. Observational data suggest that Japanese teachers use the approach of individuals completing in-class trial problems very differently from how U.S. teachers use it. In Japan this individual work is followed by planned use of the individual results in group and whole-class activities. In the U.S., this work is often not followed by any activities, only the end of the class period. We found similar results for science, although the use of in-class homework is not as pronounced.

Incremental assembly is only a hypothesis. It does appear to explain characteristic patterns of U.S. science and mathematics instruction. Fragmentation, however, is not a hypothesis but rather a fact strongly supported by data. Some common cause must repeatedly create unfocused curricula. Something replicates textbooks that present content in small bits typically with low demands. Something results in teachers repeatedly organizing instruction into small pieces typically with low demands on students. Something leads teachers to present instruction by use of several activities in each period and to place emphasis on students doing homework and seat-

Exhibit 44.

Time grade 8 mathematics teachers reported having students do exercises or homework during class.

U.S. teachers allow for far more homework in class than do the other two countries on average. All three countries make extensive use of "seatwork," (in-class exercises done by individuals but related to the current lesson). Japanese teachers report using such in-class exercises considerably more than do U.S. teachers. [This exhibit displays the average number of minutes of in-class time teachers report using for students to do exercises (as part of the current lesson) and "homework" (as follow-up to a completed lesson).]

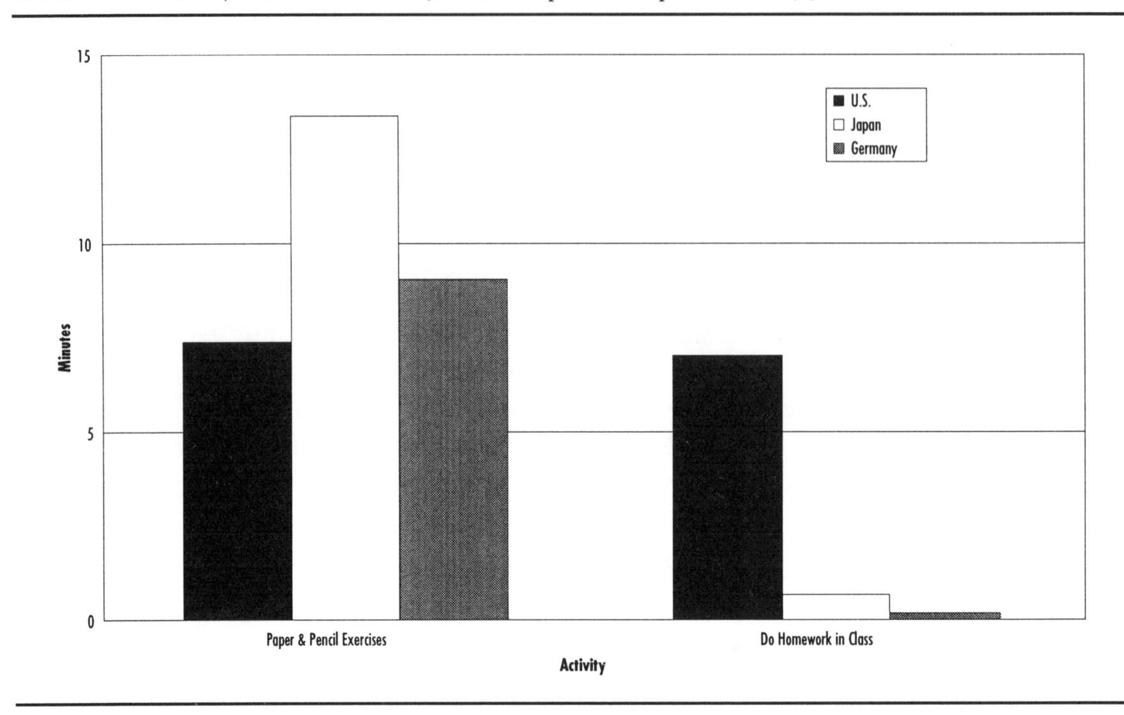

work. Incremental assembly is a sufficient explanation for these manifestations of fragmentation. However, we cannot prove that it is the necessary — the only — explanation for these data. This incremental approach has a strong basis in U.S. culture of this century and in our educational history. We believe it is a strong possibility as an explanation for the second kind of splintering found in our guiding visions for mathematics and science education.

Sources of Splintering: Summary and Concluding Remarks

How have the guiding visions for mathematics and science education become so splintered? Curricular thinking and planning has been shaped in ways that lead away from rather than follow from a coherent intellectual vision of what is and should be done. At least two kinds of causes seem involved.

Organizationally, curricular decision-making and guidance are in the hands of many "actors" in a loosely coupled arrangement of relatively independent organizations. Each appears to pursue their traditional patterns of behavior, their own perceived goals, and to act in a way appropriate for those to whom they perceive themselves as accountable. Only informal links exist among these organizations. They pursue individual visions that sometime accord and at other times conflict. From this flow composite curricula that are unfocused and inclusive in trying to provide for at least something of many recommended topics and student demands in learning science and mathematics. Commercially produced textbooks must assume a cautious, inclusive stance in the face of such diverse and, at times, conflicting curricular visions. Teachers inherit goals and resources that reflect inclusiveness and lack of focus. They respond accordingly. The loosely federated union of the many organizations that shape the mathematics and science curricula in U.S. classrooms results in a splintered vision and a fragmented reality.

At the same time, there seems to be something more than organizational complexities that cause the constant replication of fragmented, low-demand curricula that skitter among contents and activities and encourage teachers to do the same. Fragmentation is a fact in U.S. mathematics and science curricula, textbooks, and teaching. In that sense, as well as the organizational sense, our visions of what to do in helping children learn mathematics and the sciences are splintered.

One possible explanation for this is a deeply seated American ideology of a technological approach to mass science and mathematics education. This ideology may spring from our national experience of the power of industrial and assembly line production. More immediately it may spring from its psychological and educational counterparts (behavioral objectives, programmed instruction, etc.). These have been perennial parts of U.S. education. We have applied the term *incremental assembly* to this ideology and believe that it is a very likely intellectual foundation that shapes how we think about mathematics and science curricula. It may well keep us from finding other, more coherent and powerful ways to think about and organize curricula.

Chapter 5

So What Can We Expect from U.S. Students?

U.S. educational "yield" in its students' science and mathematics achievements will be obtained by a composite educational system guided by a splintered vision. This yield is shaped by unfocused curricula, cautious and inclusive textbooks, and teachers who devote small amounts of instructional time to many topics. Will our educational system's yield in science and mathematics education be what we desire or expect? Evidence from other sources suggests that student achievement in the context of unfocused expectations is likely to be low.[45] Data from Japan and other countries show that alternative, potentially more focused approaches are not impossible. In the U.S. context of many states, however, they may be more difficult.

What Are Our Basics and Are They the Same as Other Countries' Basics?

Many of a country's composite attainments are determined by what they regard as basic in science and mathematics education and how well they communicate and support those basics. What are our basics and how do they compare with those of other countries?

Population 1 mathematics. Internationally what is basic in Population 1 mathematics is mainly number ('whole numbers,' 'common and decimal fractions'), measurement ('units,' 'perimeter, area, and volume'), geometry (polygons, circles, points, lines, simple coordinate geometry, transformations) and, in textbooks only, data representation topics. (See Exhibit 11 and Exhibit 21.) The U.S. composite curriculum corresponds fairly well with this international composite. The U.S. composite includes the same number, measurement, geometry, and data representation topics. These are supplemented with additional measurement and geometry topics, simple probability, estimation, beginning algebra content in functions and equations, and, in textbooks only, proportionality concepts and several other topics.

At first glance, it seems that U.S. composite contains all of the international composite and more. This would suggest the common topics (the core) for the U.S. compare favorably with the international core of common topics. The difficulty is in the "and more." Because the U.S. gives brief treatment to so many topics, a simple listing of topics mentioned is misleading. The portrait of textbook emphases (in Exhibit 21) is more revealing. Compared to Japan, the U.S.

devotes about the same amount of space to whole number operations and common fractions. We devote considerable space to units of measurement, while Japan has moved on to focus on perimeter, area, and volume, on decimal fractions, and has begun algebraic work with number patterns, relations and functions (presumably more the former than the latter). A more detailed analysis shows our Population 1 mathematics deals with some less demanding content and makes it clear that we treat some mentioned material with misleading brevity.

Population 2 mathematics. The differences in Population 2 mathematics are more pronounced. (See Exhibit 12 and Exhibit 23). For grade 8 mathematics students not in Algebra I classes (about 75 to 80 percent of all U.S. grade 8 students), what is common and emphasized is arithmetic ('whole numbers: operations,' 'common and decimal fractions,' 'percentages'), estimation, and 'measurement: units.' None of the arithmetic topics are part of the international composite core. Instead, internationally, the focus is on integers, rational numbers, 'exponents, roots, and radicals,' and on geometry, algebra, and proportionality topics. The U.S. does include work on integers, geometry, algebra, proportionality, and data representation. The difficulty, again, is that the U.S. composite core is much more inclusive than the international composite. Geometry, algebra, and integers are thus only a few among many topics in the U.S. They play a comparatively greater role internationally. Advanced number topics (rational numbers,'exponents, roots, and radicals') are not even part of the U.S. composite.

What is most emphasized in U.S. grade 8 mathematics textbooks is quite different from what is similarly emphasized in Germany and Japan (see Exhibit 23). While algebra is among the top five topics in the U.S., it makes up, on average, only about 10 percent of the textbook space overall. This compares to much more space in Japanese and German textbooks (almost 40 percent and almost 25 percent, respectively). The other most extensive textbook topics in Germany and Japan are geometry topics, which (other than 'perimeter, area, and volume') are not even found among the top five topics in the U.S.

Clearly Germany and Japan emphasize algebra and geometry. These topics are part of what is basic. In contrast, algebra and geometry are not serious parts of what we consider basic for the majority of U.S. grade 8 students. The story is, of course, different for U.S. Algebra I students in grade 8. However, these make up only 15 to 20 percent of the mathematics students at this level.

What U.S. teachers teach is consistent with the data for curricula and textbooks in mathematics. The combination of curricula, textbooks, and teaching gives a de facto definition of what is basic in the U.S. — arithmetic, fractions, and a relatively small amount of algebra. Algebra topics are not even among the 10 most commonly taught grade 8 mathematics topics in the U.S. (see Exhibit 45), while at least one algebra topic is among the top 3 in both Japan and Germany.

Exhibit 45.

The ten topics most commonly taught in grade 8 mathematics and science.

The exhibited lists show differences in what are operationally the *de facto* "basics" among the U.S., Germany, and Japan. [This exhibit lists the ten most commonly taught mathematics topics and the ten most commonly taught science topics as reported by grade 8 teachers surveyed on how they spent their instructional time. Further international data are not yet available for release.]

MATH

U.S.	Japan	Germany
Other Numbers & Number Concepts	Geometry: Congruence & Similarity	Equations & Formulas
Number Theory & Counting	Data Representation & Analysis	Perimeter, Area, Volume
Perimeter, Area, Volume	Patterns, Relations, & Functions	2-D Geometry
Estimation & Number Sense	2-D Geometry	3-D geometry & Vectors
Percentages	Proportionality Problems	Measurement: Units
2-D Geometry	Other Content	Geometry: Congruence & Similarity
Proportionality Concepts	Estimation & Number Sense	Proportionality Problems
Proportionality Problems	Proportionality Concepts	Percentages
Properties of Common & Decimal Fractions	Measurement: Estimation & Error	Patterns, Relations, & Functions
Relationships of Common & Decimal Fractions	Equations & Formulas	Other Numbers & Number Concepts

SCIENCE

U.S.	Japan	Germany
Nature of Science	Human Diversity	Environmental & Resource Issues
Structure of Matter	Human Life Processes & Systems	Nature of Science
Matter	Chemical Transformations	Energy & Physical Processes
Science, Technology, Mathematics	Structure of Matter	History of Science & Tech.
Physical Transformations	Life Processes & Systems	Heat, Temperature, Wave & Sound
Environmental & Resource Issues	Matter	Interactions of Living Things
Energy & Physical Processes	Energy & Physical Processes	Structure of Matter
Chemical Transformations	Earth Processes	Diversity, Org, & Struc. of Living Things
Heat, Temperature, Wave & Sound	Atmosphere	Life Processes & Systems
History of Science & Tech.	Environmental & Resource Issues	Matter

Population 1 science. What is in the U.S. composite science curriculum corresponds more with the international core than was true for mathematics (see Exhibit 13). We will focus more on what textbooks include (see Exhibit 22) and what teachers teach.

At this level, Japan uses textbooks that already include topics such as the 'physical properties of matter' and the earth's physical cycles, as well as the classification topics ('plant and fungi types,' 'animal types') common to Japanese and U.S. textbooks and to the international composite core. In short, the common conceptions of what is basic in Population 1 science is mostly consistent with the international composite, but leaves out some more advanced physics topics.

Population 2 science. It is more difficult to draw conclusions for science than for mathematics because the composite core is smaller for science. The topics common across state science curricula are few (see Exhibit 14). However, this may argue that there is less agreement on what is basic in science curricula. This would cause our composite "basics" — the things most commonly included in state science curricula across the U.S. — to include fewer topics. How few there are likely will affect our aggregate achievements in science. What is in the U.S. composite science curriculum, though only relatively few topics, is somewhat consistent with the international composite. We will necessarily focus more on what textbooks include (see Exhibit 24) and what teachers teach.

U.S. textbooks at this level are typically specialized, focusing on physical sciences, earth sciences, or life sciences. U.S. grade 8 physical science textbooks do include some of the advanced topics found in the international composite science curriculum, as do life sciences textbooks. The specialized texts include such topics as the 'chemical properties of matter,' 'energy,' 'classification of matter,' and 'physical properties of matter.' However, only some U.S. students will work with any one of these texts. This restriction of what is basic at the grade level to one area of the sciences has implications for achievement. If only some of the random sample of U.S. students tested have worked with the corresponding content on the test, their results on the test as a whole will be lower. We have an interesting empirical question. If sets of students who studied only some of the content relevant to the test each do well on their area and if the sets cover most areas of the test, will the aggregate performance match that of students who studied all areas covered but necessarily studied more topics?

The average U.S. textbook (general science) at this level continues to focus on classification topics ('rocks and soils,' 'land forms,' 'bodies of water') but begins to include 'physical properties of matter' among the five most emphasized topics. However, this *de facto* conception of what is basic in the U.S. at the aggregate level and for the approximately 30 percent of U.S. students who study general science contrasts sharply with the international composite. The international composite includes advanced biology topics and many physical science topics more like those in U.S. grade 8 single area specialized texts.

While we might be tempted to dismiss these differences as matters of national (or state) preference, the fact is that there are many content topics widely common among the other TIMSS countries but not widely present in the U.S., either in state curriculum guides or textbooks. This difference will definitely affect comparative achievement assuming that the TIMSS tests reflect well what is common among the majority of countries. We can dismiss lower achievements, if they occur, by claiming a mismatch between curriculum and test. However, can we truly dismiss the effects of the achievement differences if they have some part in creating economically competitive workforces?

German and Japanese textbooks are generally more consistent with the international composite. They include 'classification of matter,' 'light,' chemical properties and changes of matter, 'dynamics of motion, electricity,' and the biology topic of organism 'energy handling' among their five most emphasized textbooks.

One bright spot for U.S. hopes in science is that U.S. teachers *do* teach many of the topics that are common and basic internationally. Among the 10 most commonly taught topics in the U.S. are 'matter' and its structure, 'physical transformations' of matter, energy topics, and 'chemical transformations.' This is somewhat consistent with the international composite and very similar to what Japanese and German teachers report teaching. The fact that U.S. textbooks at this grade are specialized for most students, i.e., earth, life or physical science, but yet include a very large number of topics, many in other areas of science (e.g. earth science topics in the physical science textbook), partly explains this result. At this grade level, science as delivered by teachers in classroom instruction is more consistent with the international composite than was the case for mathematics (see Exhibit 45). We may expect science achievements to be somewhat better.

Who Are Our Curricular Peers?

Being able to compete economically, either for individuals or for nations, is not the only reason for sound educations in mathematics and the sciences. Certainly knowledge, skills and abilities are rewarding for individuals and contribute to their quality of life. They both help satisfy innate curiosity and enhance our sense of autonomy. Aiding efforts towards this growth of individuals is part of "the pursuit of happiness" and seeking the common good. However, in a national context of competing resources for various social goods, fostering this kind of growth alone might not widely be regarded as a sufficient reason for our national, state, and local expenditures on science and mathematics education.

Literate, effective citizens are needed to thrive as a nation. Part of this literacy is being quantitatively and scientifically literate. The U.S. has never been able to afford an uninformed or unthinking citizenry. Certainly the demands of citizenship are sufficient to justify expenditures for some mathematics and science education to produce quantitatively and scientifically literate citizens. However, even this may be insufficient justification for the current levels of expenditure and certainly not for increased expenditure for mathematics and science education.

Matters of individual empowerment and maintaining an informed citizenry in a technological age may be contributing reasons for national, state, and local emphases and expenditures on mathematics and science education. However, maintaining state and national economic competitiveness seems consistently to be a basis that drives the expenditures and emphases needed for mathematics and science education in our context of competing social needs. Economic competitiveness is at least one pressing basis for establishing intellectually coherent visions and working for a state- and local-level consensus about mathematics and science curriculum policies, for deeper and more effective systemic change and reform.

We should ask, in this context, are we succeeding in maintaining our national competitiveness and preparing to continue to do so? We have already seen that what are "the basics" in U.S. mathematics and science curricula, textbooks, and teaching are often not the same as those of corresponding international composites. In matters of what is basic in teaching children mathematics and science, we are not peers with the composite of other TIMSS countries.

However, if we cannot be peers with the whole set of TIMSS countries, with whom do we find ourselves peers? That is, examining only our current composite curricula in mathematics and the sciences, which other countries have similar composite curricula? The link from curricula through textbooks and teaching to students' educational attainments is not undeniably established. However, the link of what is taught to what is learned is both logical and supported by considerable educational evidence. Thus, if we wish to know how U.S. students will do in comparative assessments of mathematics and science attainments, we gain valuable insight by examining national composite curricula (and other "inputs" to educational yield) for advanced warnings of comparative achievement.

If we take seriously that curriculum, as set forth in state guidelines and textbooks, at least sets bounds on what is broadly achieved by those taught, we should investigate who our peers are — those countries that set similar bounds to their students' achievement by the nature of their composite or national curricula in mathematics and the sciences. We can assess this preliminarily by using statistical clustering methods to identify which countries would fall together as more closely "peers" on the basis of their curricular content. For the present we can report this only for grade 8 science and mathematics (see Exhibit 46 and Exhibit 47).

In grade 8 mathematics, the U.S. composite curriculum as represented by textbooks is most like those of Australia, New Zealand, Canada, Italy, Belgium (French language system), Thailand, Norway, Hong Kong, Ireland and Iceland. In grade 8 science, we are most like New Zealand, Iceland, Greece, Bulgaria, and the Peoples' Republic of China. While the curriculum of any country is interesting and has some important features, we must ask if these are the countries with whom we are and will be trying to compete economically?

Exhibit 46.

Who are our closer peers in grade 8 mathematics?

Differences among country groups are created by a variety of topics (look for those in double digits). The cluster containing the U.S. appears to do something of everything and not too much of any one thing -the only double digit topic is algebraic equations. [The data in this exhibit are the percentages of textbook blocks devoted to framework topics in grade 8 mathematics textbooks across countries. We used them to cluster the TIMSS countries into six groups of "near peers" in terms of composite grade 8 mathematics textbook content. For each cluster, we report the average percentages of textbook blocks devoted to the specific mathematics framework topics.]

	Country Groups					
Mathematics Topics	**Australia, Belgium (Fr)*, Canada, Hong Kong, Iceland, Ireland, Italy, New Zealand, Norway, Thailand, U.S.**	**Austria, Cyprus, Greece, Hungary, Romania**	**Belgium (Fl)*, France, Switzerland**	**Scotland, Sweden, Tunisia**	**Colombia, Iran, Mexico, Philippines, Russian Federation, South Africa, Spain**	**Bulgaria, China, People's Republic of, Czech Republic, Slovak Republic, Israel, Japan, Korea, Netherlands**, Portugal, Singapore**
Numbers						
Whole Number						
Meaning	2	0	1	8	1	0
Operations	5	1	6	24	2	1
Fractions and Decimals						
Common Fractions	6	1	4	5	5	1
Decimal Fractions	5	0	4	5	3	1
Relationships of Common and Decimal Fractions	2	1	5	1	3	1
Percentages	6	2	3	5	1	1
Properties of Common and Decimal Fractions	1	0	4	0	1	0
Integer, Rational and Real Numbers						
Integers and Their Properties	5	2	2	2	9	1
Rational Numbers and Their Properties	1	4	2	1	8	1
Real Numbers, Their Subsets and Their Properties	1	6	1	0	2	1
Other Numbers and Number Concepts						
Exponents, Roots and Radicals	3	6	4	0	5	6
Measurement						
Units	5	2	8	22	1	2
Perimeter, Area and Volume	7	13	6	7	3	8
Geometry: Position, Visualization and Shape						
2-D Geometry: Coordinate Geometry	4	5	2	5	5	4
2-D Geometry: Basics	6	2	10	7	6	8
2-D Geometry: Polygons and Circles	7	12	11	4	11	21
3-D Geometry	2	24	5	3	1	3
Vectors	0	1	1	0	4	1
Geometry: Symmetry, Congruence and Similarity						
Transformations	5	5	7	1	5	3
Congruence and Similarity	1	6	3	0	2	7
Constructions using Straight-edge and Compass	1	1	3	0	2	2
Proportionality						
Proportionality Problems	2	4	4	5	2	4
Slope and Trigonometry	1	2	1	0	1	3

Exhibit 46. (cont'd)

Who are our closer peers in grade 8 mathematics?

	Country Groups					
Mathematics Topics	**Australia, Belgium (Fr)*, Canada, Hong Kong, Iceland, Ireland, Italy, New Zealand, Norway, Thailand, U.S.**	**Austria, Cyprus, Greece, Hungary, Romania**	**Belgium (Fl)*, France, Switzerland**	**Scotland, Sweden, Tunisia**	**Colombia, Iran, Mexico, Philippines, Russian Federation, South Africa, Spain**	**Bulgaria, China, People's Republic of, Czech Republic, Slovak Republic, Israel, Japan, Korea, Netherlands**, Portugal, Singapore**
Functions, Relations, and Equations						
Patterns, Relations, and Functions	3	9	2	2	8	11
Equations & Formulas	14	16	8	3	17	32
Data Representation, Probability, and Statistics						
Data Representation and Analysis	6	3	3	6	1	3
Validation and Structure						
Validation and Justification	1	6	0	0	1	2
Structure and Abstracting	2	2	2	0	0	1

* The national Research Coordinators of Belgium have only collected data from curriculum guides . Due to the great level of detail of the guides, and their extensive use, data from these are compared in this display with the textbook data supplied from all other countries.

** Netherlands' sample did not meet the 50% market coverage criterion for populations 1 and 2.

Note: Countries not included in table - Argentina, Dominican Republic, Germany, Hong Kong, Slovenia

Exhibit 47.

Who are our closer peers in grade 8 science?

Strong emphases are fewer in science than in mathematics. The cluster containing the U.S. again appears to be defined by inclusiveness and a lack of emphases. [The data in this exhibit are the percentages of textbook blocks devoted to framework topics in grade 8 science textbooks across countries. We used them to cluster the TIMSS countries into seven groups of "near peers" in terms of composite grade 8 science textbook content. For each cluster, we report the average percentages of textbook blocks devoted to the most specific science framework topics.]

	Cluster Group							
Science Topics	**Argentina Cyprus Latvia Philippines**	**Canada Mexico Netherlands Norway Spain**	**Colombia Tunisia**	**Bulgaria People's Rep. of China Greece Iceland New Zealand U.S.**	**Austria Belgium (Fr)* Hong Kong Hungary Italy Portugal Singapore Sweden**	**Iran South Africa**	**Australia Belgium (Fl)* Ireland Lithuania Russian Federation Scotland Switzerland**	**Czech Republic France Japan Korea Romania Slovak Republic Slovenia**
Earth Science								
Earth Processes								
Weather & Climate	1	6	2	2	2	12	1	3
Earth in the Universe								
Earth in the Solar System	2	2	1	2	2	7	1	0
Life Sciences								
Diversity, Organization, Structure of Living Things								
Animals	0	2	4	7	4	0	12	2
Organs, Tissues	3	5	5	11	6	2	9	3
Life Processes and Systems Enabling Life Functions								
Energy Handling	3	3	11	4	4	0	4	3
Life Spirals, Genetic Continuity, Diversity								
Life cycles	1	1	13	3	0	9	3	0
Interactions of Living Things								
Reproduction	3	2	16	2	2	10	2	2
Animal Behavior	0	1	8	2	1	0	1	1
Human Biology & Health	1	3	0	11	4	0	9	4
Physical Sciences								
Matter								
Classification of Matter	2	5	8	2	4	3	6	4
Energy & Physical Processes								
Energy Types, Sources, Conversions	14	4	2	6	3	8	5	3
Heat & Temperature	6	3	1	2	4	0	6	1
Light	7	1	0	3	6	5	4	2
Electricity	0	1	0	1	8	9	10	23
Physical Transformations								
Physical Changes	3	3	6	4	1	4	2	1
Chemical Transformations								
Chemical Changes	2	3	1	3	4	3	3	8
Forces & Motion								
Time, Space, Motion	4	3	2	3	2	0	1	1
Dynamics of Motion	3	1	2	1	1	4	1	0
Science, Technology, & Mathematics								
Interactions of Science, Mathematics, & Technology								
Applications of Science in Mathematics, Technology	6	2	0	5	4	0	3	3
Environmental & Resource Issues								
Food Production, Storage	0	2	8	2	1	0	2	1

* The national Research Coordinators of Belgium have collected data only from curriculum guides. Due to the great level of detail of the guides, and their extensive use, data from these are compared in this display with the textbook data supplied from all other countries.

Note: Countries not included: Denmark, Dominican Republic, Germany (did not code Biology text) and Israel.

Are These the Peers We Want?

Are the curricular peers we have the peers we want? Certainly we desire to empower and inform our citizenry comparably as well as to effectively compete economically with other developed countries. We want attainments similar to the European Union, to the APAC countries (especially Japan and the "young tigers" of Korea, Singapore, etc., countries together in a quite different cluster), and, most definitely, with the other G-7 countries (how many G-7 countries do we find as our close curricular peers?).

Mathematics and science education clearly should not be shaped solely by concerns of economic competition. Even were that to be so, cultural differences would lead to different emphases in defining what is most powerful and appropriate in mathematics and science education for each culture. However, when we find ourselves most similar to countries other than those with whom we seek to be peers, we have reason for deep concern. Certainly further and more sophisticated analyses are in order. We do not, however, want to become victims of the particular sophistry that "Further research will always support our own position." Considering the data thus far, we as a nation must be concerned.

The analyses reported here deal only with grade 8 science and mathematics. What can we conclude from them? We have a highly fragmented curriculum in mathematics, textbooks that are a "mile wide and an inch deep," teachers that cover many topics, none extensively, textbooks with low demands on students and a much more limited conception compared to the international notion of what are "the basics." It seems highly likely that U.S. student achievement in mathematics will be below international averages. It seems certain to be below that of many of our current economic peers.

Our science curriculum is less fragmented, the textbooks cover many more topics than their international counterparts, but this is somewhat modified by textbooks with greater focus in specific areas such as earth, life and the physical sciences, teachers cover more of what is consistent with some of our economic peers, and what is basic seems more comparable to the international composite. However, this focus and those basics are achieved by distributing topics to segments of our student population. Further, the average demand of textbooks is low (although this is also true in other TIMSS countries). Science achievement seems likely to be closer to international averages, but still not what we desire and certainly below some, if not most, of our current economic peers. Comforts are few and the causes for concern are many in science as well as mathematics education.

Chapter 6

OPEN QUESTIONS: HOW DO WE GET WHERE WE WANT TO GO?

We can summarize in one sentence our findings about U.S. mathematics and science education: we as a nation are not where we want to be. Vision is lacking that should guide what we do to provide U.S. children educations in mathematics and the sciences. There is no single, coherent, intellectual vision underlying our efforts. Reform efforts in science and mathematics education offer such powerful visions, but these visions take time to reach local classrooms and do so only after state and local interpretation and adaptation. Our guiding visions for science and mathematics are splintered.

In part, the U.S. vision necessarily lacks coherence because the rights of many independent local, state, and national organizations lead them to pursue their own visions of science and mathematics education. This loose federation of organizations carries out aggregate national educational efforts. Some of the main support for instruction is provided by commercial textbook publishers who interact with but are independent from these organizations. The voices of professional and other organizations making recommendations and calling for change must address not one audience but these many audiences. The conversations shaping guiding visions of mathematics and science education in the U.S. are less a conversation and more a babel. While independence is essential, the result in this case is a composite of sometimes corresponding, sometimes conflicting separate visions. This collection of organizations and multiple voices is inescapable in shaping U.S. educational policy. We hope the splintering is not.

In part, the U.S. vision is also fragmented because it views school mathematics and science as partitioned into many topics that form the building blocks of curricula. The visions that shape those fragmented ideas, concepts, and skills into integrated curricula are often weighted by tradition, inertia, false assumptions, and the lack of an articulated intellectual conception that would help these pieces coalesce into focused, goal-directed action.

These separate science and mathematics curricula do have goals. They have, if anything, too many goals. The curricula in mathematics and the sciences, both state and the national composite, are unfocused and inclusive. They proceed by accumulating newer topics (e.g., reform recommendations) without eliminating others, and seem to involve no strategic concept of focusing on fewer key goals, linking contents, and setting higher demands on students. Through inertia, caution, or imitation with less insight than desirable, our curricula perpetuate fragmentation.

U.S. textbooks are published by those who must be commercially viable in an arena of often conflicting demands. There are conflicting opinions about what is basic, which contents to include, what demands to set on students, and how to organize materials to better support coherent classroom instruction. For commercial viability, cautious visions typically guide these publishers. The results are textbooks that often have basic content not as advanced as that in comparable grades in other countries. They have far more content topics (since virtually all suggestions have been incorporated). However, because they are so inclusive, they typically allocate little space to most topics and provide comparatively more materials on less demanding student tasks.

U.S. teachers do what we ask them to do. We hand them fragmented, inclusive curricula — curricula "a mile wide and an inch deep." We provide as their fundamental resources encyclopedic, inclusive textbooks filled with brief coverage of many topics and with a preponderance of low-demand tasks for students. Textbooks that are organized to skitter continually among topics. Perhaps this constantly changing micro-organization is in aid of maintaining student interests or, nominally, of making connections among topics (no one of which has typically been covered sufficiently to anchor any connection).

U.S. teachers most often face demanding workplaces. For survival, or perhaps imbibing the spirit of the curricula and textbooks delivered to them, they organize sequences of lessons briefly covering most topics. They typically make few serious demands on their students and move rapidly among activities. Their beliefs suggest they might choose differently under other circumstances. They are informed — not thoroughly but sufficiently to suggest that more information is not viable as a single approach to our problems. They are simply placed in situations in which they cannot do their best. Our teachers respond to the demands communicated to them. If those demands were more focused and informed, and conditions changed to allow it, our teachers might well respond to those more powerful demands.

The reasons are not clear for the current state of our nation's composite visions guiding mathematics and science education. We continue to replicate fragmented, low-demand curricula in mathematics and the sciences. Organizationally, our federation is essential and has its historical roots in our national structure. However, we find it less clear why, in each of our many, mainly independent, sites for curricula and policy, we repeatedly replicate fragmentation and its concomitants in curricula, textbooks, and teaching. Perhaps this action is deeply rooted in impacts of American traditions of mass production on our national efforts at mass education.

The assembly line predates efforts at extensive mass public education in an expanding nation.[46] Extensive mass production (starting, perhaps, with Henry Ford's first automobile assembly line in 1913) certainly predates mass *secondary* education. The idea of producing uniform, interchangeable parts that can be assembled into desired wholes is a concept that generalizes from the history of mass production. It also has roots in behavioral psychology. It has entered education at least as behavioral objectives and programmed instruction, and may seem an ideal strategy for our highly mobile society. This notion may be sufficient to explain our curricula of many small topics, frequent low demands, interchangeable pieces of learning to be assembled later, persisting coverage, and distributed mastery. U.S. history makes it logical. The current state of mathematics and science education that may reflect these notions — in unfocused curricula, inclusive textbooks, teachers faced with many topics — is unfortunate. Henry Ford, presumably, did not try to make all models simultaneously on the same assembly line.

What of the nation's students? Will they be able to compete effectively internationally in mathematics and the sciences? With the "playing floor" we have given them, successful competition would be extraordinary. Our students at times are extraordinary, as are our teachers, but that is not something that can be assumed in setting educational policy. We have embodied in our "basics" contents and goals that are not as demanding or well organized as those of many other countries. We have produced textbooks that do not yet present material best designed to produce the highest student achievement. This is true especially for more complex tasks and for those requiring integrated knowledge of a content. Our close curricular peers contain few, if any, of those with whom we now compete economically and with whom we wish to continue to compete effectively.

U.S. students' achievement results — the yield of our aggregate national education "system" — in mathematics and the sciences are likely to be disappointing and not all through our students' fault. We must make changes if we are to compete and to produce a quantitatively, scientifically literate, adequately proficient workforce and citizenry.

The authors of this report do not represent any official or policy-making group. Our job has been to design relevant research, analyze its results carefully, and report them objectively. We have tried to do this here and elsewhere. Because of who we are, we do not feel it appropriate to make specific recommendations. Because of what we have done, we have opinions that we think are informed by data. There are conclusions we can draw that seem to us correct and, if we cannot make recommendations, we can at least ask questions — questions that our results lead us to believe important for those who do set policy.

We as a nation are not where we want to be. So, how do we get there? The precise route may be unclear but some milestones along the route are quite clear. We outline those milestones by posing questions. Most of these questions are not original with us; although, their form here has been influenced by the data we investigated. In fact, some efforts are currently underway to address these questions including the National Science Foundation's State Systemic Initiatives and the recently convened executive committee of the National Governor's Association in conjunction with business leaders. Those milestones include answers to several pressing questions:

- *How can we gain — an intellectually coherent vision — focus in our mathematics and science curricula and textbooks?* Working with the real U.S. national education system of a federation of organizations and a conversation of many voices, and given the role of commercial publishers, how can we empower our curricula and re-shape our textbooks to support our teachers in carrying out those powerful new visions?

- *How can we raise expectations and demands on our students?* While it is inappropriate to increase demands, on students or teachers, without increasing their chances to meet those raised expectations, how can we provide a reasonable climate of higher expectations? Will providing strategic, focused curricula and supporting textbooks naturally raise expectations or will special efforts be needed?

- *How can we help our teachers to do the best they can in teaching mathematics and sciences to our students?* Will better focused, more coherent curricula suffice? Will they suffice if supported by more appropriately organized textbooks? Will a more focused and coherent curriculum permit correspondingly more focused teacher in-service? Would such focus permit better teacher preparation in the subject-matter areas?

- *Can we find a better model for curriculum and instruction?* If distributed mastery does not work, what can replace it? If incremental assembly leads to fragmentation rather than efficient reproduction, should it not be examined and questioned more closely? What curricular ideologies or, better, what coherent visions of how education is to take place can we find as a source for better practical models to shape curriculum and instruction?

- *Can we develop a new vision of what is basic and important?* Even if we remain somewhat splintered and fragmented, can we begin to be less inclusive and more focused? Can we use what we learn about the successful distribution of educational opportunities in other countries to gain a new vision of what is basic and what is not for U.S. science and mathematics education? Should not our national vision of what is basic be at least as high as the common expectations in the rest of the TIMSS countries?

These questions mark the way for public discussion seeking intellectually coherent underpinnings and guiding concepts to reshape our national production of mathematics and science education. The details will vary. The vision for science will not be the same as the vision for mathematics. However, in both cases, we wish our teachers' classroom activities not to be replays of disjoint episodes. We want our teachers to carry out their instruction in a way that gradually unfolds a coherent story. Clearly, however, our teachers cannot tell a coherent story if we do not have one. It is this kind of deeply directive, foundational vision that we must find. Only this may integrate the many details of mathematics and science education to serve central purposes and powerful, reachable goals worth pursuing.

Raising these questions does not presuppose specific answers. The strategies that emerge for producing additional change may be quite varied. Coalitions of states and local school districts may voluntarily join to define an intellectually coherent vision for mathematics and science education. Textbook publishers may join states and professional organizations in reshaping and defining such visions. Consensus building and the continued impact of mathematics and science reform efforts may produce more extensive, consistent, and focused strategies for change. A viable role for federal representation in our national interests may yet be found. Some combination of all of the above, or some completely unforeseen strategy may emerge to help bring needed changes. What we seek to do in raising these questions is to question complacency about the current state of mathematics and science education, to present data from a relevant empirical investigation of their current state and to begin discussion by helping bring to light questions that seem to us implied by that current state.

What kinds of mathematics and science education do we as a nation want for our children? While this is a central question for public debate, it seems likely that we want educations that

- are more focused, especially on powerful, central ideas and capacities;
- provide more depth in at least some areas, so that the content has a better chance to be meaningful, organized, linked firmly to children's other ideas, and to produce insight and intuition rather than rote performance; and
- provide rigorous, powerful, and meaningful content that is likely to produce learning that lasts and not just learning that suffices for the demands of schooling.

We need this kind of powerful mathematics and science education because they

- provide a strong basis for our democracy by helping create a literate and informed citizenry;
- help each individual to grow, develop, come nearer to reaching their individual potential, and to feel more autonomous and empowered in areas that have been sources of frustration and anxiety for many; and
- provide a sound basis for continuing national prosperity, even if this must be achieved in a competitive, information driven, technological, and changing international arena.

Certainly these lists must be modified and expanded in our collective discussion of educational futures but they at least illustrate something of the range of concerns appropriate to shaping goals for mathematics and science education.

The TIMSS data are resources in a time calling for change and needing all the resources it can get. We have shown something of what our data reveal about the current state of U.S. science and mathematics education. We have offered more technical and detailed reports and will do so again. Here we have presented some central data essential to understanding the current state of U.S. mathematics and science education and have raised questions we think important.

Certainly these are not the only questions that must be asked and answered on the way to the revolution or, if one prefers, on the way to a fruitful evolution in mathematics and science education. We have not touched on whole domains of issues — for example, concerns for equity in educational opportunity. We have done this not because we are unaware of other issues or because our data are irrelevant to all of them. Rather, we have done this because in arguing for focusing rather than splintering we wished ourselves to be focused. Others must join in seeking answers to the questions raised here and the others we did not raise. Our data can help. Presently, however, our story is simple. The U.S. vision of mathematics and science education is splintered. We are not where we want to be. We must change.

Appendix A

TIMSS CURRICULUM FRAMEWORKS: MEASURING CURRICULAR ELEMENTS

The Third International Mathematics and Science Study (TIMSS) was designed to involve extensive data collections from participating national centers, an analysis of curriculum documents, a range of questionnaires, and complex attainment testing. This multi-component data collection (carried out in participating countries and at several international sites) made it essential that all TIMSS components be linked by a common category framework and descriptive language. Whether classifying a test item, characterizing part of a curriculum document, or linking a questionnaire item to other TIMSS parts, any description had to use common terms, categories, and standardized procedures to assign numerical codes for entry into the appropriate database.

This common language was provided by two framework documents — one for the sciences and one for mathematics. Each covered the full range of the years of schooling in a unified category system. Each framework was articulated first only in technical reports (as is still true for more extensive explanatory notes) but has now been reproduced and documented in a monograph.*

Each framework document was multi-faceted and multi-layered. It considered three aspects of subject matter content and performance — *content* (subject matter topic), *performance expectation* (what students were expected to do with particular content), and *perspective* (any over-arching orientation to the subject matter and its place among the disciplines and in the everyday world). Content as subject matter topic is relatively clear; "performance expectation" is less so. It was decided not to attempt identifying student (cognitive) performance processes — which would be highly inferential, particularly subject to cultural differences, and thus not feasible in a cross-cultural context. It was decided rather to specify expectations for student (science and mathematics) performances — for example, formulating or clarifying a problem to be solved, developing a solution strategy, verifying a problem solution, and so on, without positing cognitive means for producing the performances. This focused on more generic, less

*Robitaille, D. F. and others, Curriculum Frameworks for Mathematics and Science, (Vancouver: Pacific Educational Press, 1993).

culture-bound task expectations and demands. Postulating specific cognitive processes necessarily would have involved culturally tied cognitive categories inherent in student thinking in any given country.

In addition to being a multi-aspect system, each framework was designed for using multiple categories. Each element — curriculum guide, textbook segment, test item — could be considered as part of more than one category of any framework aspect. Each element was to be classified in as many framework categories as needed to capture its richness. Each would have a unique, often complex "signature" — a set of content, performance expectation, and perspective categories that characterized it, at least in terms of the framework's three aspects. This system is flexible — allowing simple or more complex signatures as needed.

Such flexible multidimensionality was essential for analyzing curriculum documents. How often segments involved multiple contents or performance expectations had to be determined empirically by the documents, not by the study design. Characterizing mathematics and science curricula required a tool permitting coherent categorization of curriculum guides' and textbooks' major pedagogical features. It had to be capable of translating many nationally idiosyncratic ways for specifying mathematics and science education goals into a common specification language. The TIMSS mathematics and science curriculum frameworks were designed to be such a tool.

Framework development was cross-national. The frameworks had to be suitable for all participating countries and educational systems. Representing the interests of many countries, the frameworks were designed cross-nationally and passed through several iterations. The result is imperfect, but it is still a step forward in cross-national comparisons of curricular documents. Its value may be judged by the results of this and related volumes.

Each framework aspect is organized hierarchically using nested subcategories of increasing specificity. Within a given level, the arrangement of topics does not reflect a particular rational ordering of the content. See Figure A.1 for an overview of the content aspect of the science framework. The end of this appendix briefly presents all three aspects of the mathematics and science frameworks. Each framework was meant to be encyclopedic — intended to cover all possibilities at some level of specificity. No claim is made that the "grain size" — the level of specificity for each aspect's categories— is the same throughout the framework. Some subcategories are more inclusive and commonly used, others less so. Specificity had broad commonalties but also considerable variation among the participating countries. This varying granularity requires special care in designing framework-based methods and interpreting the results of their use.

Figure A.1

Content Categories of the Science Framework.

Each aspect of the framework contained a set of main categories. Each main category contained one or more levels of more specific subcategories. The main content categories are shown here with some subcategories expanded to give a better insight into the framework's structure.

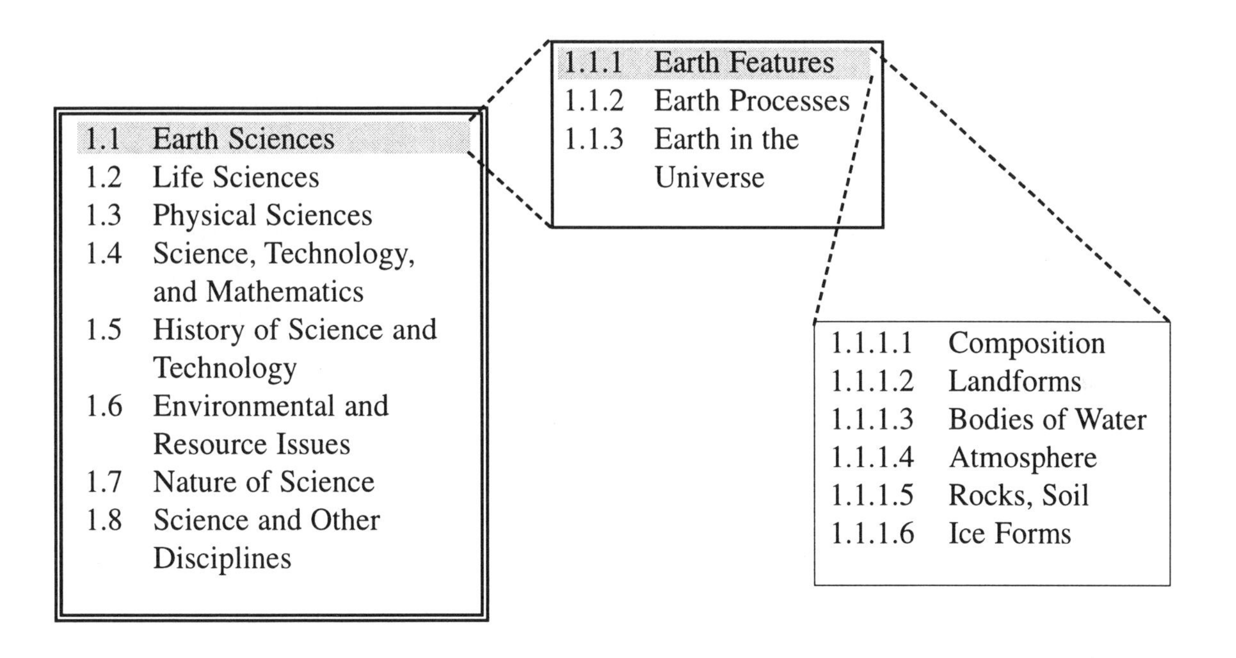

In the mathematics framework, *content* involves 10 major categories, each with 2 to 17 subcategories. Some subcategories are divided still further. The level of detail and organization reflects a compromise between simplicity (fewer categories) and specificity (more categories). The hierarchical levels of increasing specificity allow some flexibility in detail level and generalization.

In the science framework, *content* involves eight major categories, each with two to six subcategories. Some subcategories are divided further. The level of detail and organization reflects compromise between simplicity (fewer categories) and specificity (more categories). The hierarchical levels of increasing specificity allow some flexibility in detail level and generalization.

Performance expectations is a flexible category system (see below for details), but — as with the other framework aspects — no category or subcategory is considered exclusive. Any document segment should involve at least one or more performance expectation categories from the framework. Complex, integrated performances can thus be characterized in detail — as can contents and perspectives. This differs from traditional grid classifications that generate unique categorizations combining one element from two or more dimensions.

Complex signatures reveal important differences in how curricula are meant to achieve their goals. They show differences in how subject matter elements are combined — and differences in what students are expected to do. Each framework can reveal subject matter presented in an integrated, thematic way, with a rich set of performance expectations for students as recommended by curriculum reformers in many countries. However, it also allows simpler "signatures" — for example, those often associated with more traditional curricula and many traditional achievement test items.

What follows is a listing of the content, performance expectation, and perspective codes of the mathematics framework. For a more detailed discussion, see the book referenced in footnote 1 of this appendix or the technical report of "Explanatory Notes" for the mathematics framework.

Content

1.1 Numbers

- 1.1.1 Whole numbers
  - 1.1.1.1 Meaning
  - 1.1.1.2 Operations
  - 1.1.1.3 Properties of operations
- 1.1.2 Fractions and decimals
  - 1.1.2.1 Common fractions

1.1.2.2 Decimal fractions
1.1.2.3 Relationships of common and decimal fractions
1.1.2.4 Percentages
1.1.2.5 Properties of common and decimal fractions

1.1.3 Integer, rational, and real numbers
1.1.3.1 Negative numbers, integers, and their properties
1.1.3.2 Rational numbers and their properties
1.1.3.3 Real numbers, their subsets, and their properties

1.1.4 Other numbers and number concepts
1.1.4.1 Binary arithmetic and/or other number bases
1.1.4.2 Exponents, roots, and radicals
1.1.4.3 Complex numbers and their properties
1.1.4.4 Number theory
1.1.4.5 Counting

1.1.5 Estimation and number sense
1.1.5.1 Estimating quantity and size
1.1.5.2 Rounding and significant figures
1.1.5.3 Estimating computations
1.1.5.4 Exponents and orders of magnitude

1.2 Measurement

1.2.1 Units
1.2.2 Perimeter, area, and volume
1.2.3 Estimation and errors

1.3 Geometry: Position, visualization, and shape

1.3.1 Two-dimensional geometry: Coordinate geometry
1.3.2 Two-dimensional geometry: Basics
1.3.3 Two-dimensional geometry: Polygons and circles
1.3.4 Three-dimensional geometry
1.3.5 Vectors

1.4 Geometry: Symmetry, congruence, and similarity

1.4.1 Transformations
1.4.2 Congruence and similarity
1.4.3 Constructions using straight-edge and compass

1.5 Proportionality

1.5.1 Proportionality concepts
1.5.2 Proportionality problems

1.5.3 Slope and trigonometry
1.5.4 Linear interpolation and extrapolation

1.6 Functions, relations, and equations
1.6.1 Patterns, relations, and functions
1.6.2 Equations and formulas

1.7 Data representation, probability, and statistics
1.7.1 Data representation and analysis
1.7.2 Uncertainty and probability

1.8 Elementary analysis
1.8.1 Infinite processes
1.8.2 Change

1.9 Validation and structure
1.9.1 Validation and justification
1.9.2 Structuring and abstracting

1.10 Other content
1.10.1 Informatics

Performance Expectations

2.1 Knowing
2.1.1 Representing
2.1.2 Recognizing equivalents
2.1.3 Recalling mathematical objects and properties

2.2 Using routine procedures
2.2.1 Using equipment
2.2.2 Performing routine procedures
2.2.3 Using more complex procedures

2.3 Investigating and problem solving
2.3.1 Formulating and clarifying problems and situations
2.3.2 Developing strategy
2.3.3 Solving
2.3.4 Predicting
2.3.5 Verifying

2.4 Mathematical reasoning
- 2.4.1 Developing notation and vocabulary
- 2.4.2 Developing algorithms
- 2.4.3 Generalizing
- 2.4.4 Conjecturing
- 2.4.5 Justifying and proving
- 2.4.6 Axiomatizing

2.5 Communicating
- 2.5.1 Using vocabulary and notation
- 2.5.2 Relating representations
- 2.5.3 Describing/discussing
- 2.5.4 Critiquing

Perspectives

3.1 Attitudes toward science, mathematics, and technology

3.2 Careers involving science, mathematics and technology
- 3.2.1 Promoting careers in science, mathematics, and technology
- 3.2.2 Promoting the importance of science, mathematics, and technology in nontechnical careers

3.3 Participation in science and mathematics by underrepresented groups

3.4 Science, mathematics, and technology to increase interest

3.5 Scientific and mathematical habits of mind

The following is a brief listing of the content, performance expectation, and perspective codes of the science framework. For a more detailed discussion see the monograph referenced or the technical report of "Explanatory Notes" for the science framework.

1.1 Earth sciences

1.1.1 Earth features

1.1.1.1 Composition

1.1.1.2 Landforms

1.1.1.3 Bodies of water

1.1.1.4 Atmosphere

1.1.1.5 Rocks, soil

1.1.1.6 Ice forms

1.1.2 Earth processes

1.1.2.1 Weather and climate

1.1.2.2 Physical cycles

1.1.2.3 Building and breaking

1.1.2.4 Earth's history

1.1.3 Earth in the universe

1.1.3.1 Earth in the solar system

1.1.3.2 Planets in the solar system

1.1.3.3 Beyond the solar system

1.1.3.4 Evolution of the universe

1.2 Life sciences

1.2.1 Diversity, organization, structure of living things

1.2.1.1 Plants, fungi

1.2.1.2 Animals

1.2.1.3 Other organisms

1.2.1.4 Organs, tissues

1.2.1.5 Cells

1.2.2 Life processes and systems enabling life functions

1.2.2.1 Energy handling

1.2.2.2 Sensing and responding

1.2.2.3 Biochemical processes in cells

1.2.3 Life spirals, genetic continuity, diversity

1.2.3.1 Life cycles

1.2.3.2 Reproduction

1.2.3.3 Variation and inheritance

1.2.3.4 Evolution, speciation, diversity

1.2.3.5 Biochemistry of genetics

1.2.4 Interactions of living things

1.2.4.1 Biomes and ecosystems

1.2.4.2 Habitats and niches

1.2.4.3 Interdependence of life

1.2.4.4 Animal behavior

1.2.5 Human biology and health

1.2.5.1 Nutrition

1.2.5.2 Disease

1.3 Physical sciences

1.3.1 Matter

1.3.1.1 Classification of matter

1.3.1.2 Physical properties

1.3.1.3 Chemical properties

1.3.2 Structure of matter

1.3.2.1 Atoms, ions, molecules

1.3.2.2 Macromolecules, crystals

1.3.2.3 Subatomic particles

1.3.3 Energy and physical processes

1.3.3.1 Energy types, sources, conversions

1.3.3.2 Heat and temperature

1.3.3.3 Wave phenomena

1.3.3.4 Sound and vibration

1.3.3.5 Light

1.3.3.6 Electricity

1.3.3.7 Magnetism

1.3.4 Physical transformations

1.3.4.1 Physical changes

1.3.4.2 Explanations of physical changes

1.3.4.3 Kinetic theory

1.3.4.4 Quantum theory and fundamental particles

1.3.5 Chemical transformations

1.3.5.1 Chemical changes

1.3.5.2 Explanations of chemical changes

1.3.5.3 Rate of change and equilibria

1.3.5.4 Energy and chemical change

1.3.5.5 Organic and biochemical changes

1.3.5.6 Nuclear chemistry

1.3.5.7 Electrochemistry

1.3.6 Forces and motion

1.3.6.1 Types of forces

1.3.6.2 Time, space, and motion

1.3.6.3 Dynamics of motion

1.3.6.4 Relativity theory

1.3.6.5 Fluid behaviour

1.4 Science, technology, and mathematics

1.4.1 Nature or conceptions of technology

1.4.2 Interactions of science, mathematics, and technology

1.4.2.1 Influence of mathematics, technology in science

1.4.2.2 Applications of science in mathematics, technology

1.4.3 Interactions of science, technology, and society

1.4.3.1 Influence of science, technology on society

1.4.3.2 Influence of society on science, technology

1.5 History of science and technology

1.6 Environmental and resource issues related to science

1.6.1 Pollution

1.6.2 Conservation of land, water, and sea resources

1.6.3 Conservation of material and energy resources

1.6.4 World population

1.6.5 Food production, storage

1.6.6 Effects of natural disasters

1.7 Nature of science

1.7.l Nature of scientific knowledge

1.7.2 The scientific enterprise

1.8 Science and other disciplines

1.8.1 Science and mathematics

1.8.2 Science and other disciplines

Performance expectations

2.1 Understanding

2.1.1 Simple information

2.1.2 Complex information

2.1.3 Thematic information

2.2 Theorizing, analyzing, and solving problems

- **2.2.1 Abstracting and deducing scientific principles**
- **2.2.2 Applying scientific principles to solve quantitative problems**
- **2.2.3 Applying scientific principles to develop explanations**
- **2.2.4 Constructing, interpreting, and applying models**
- **2.2.5 Making decisions**

2.3 Using tools, routine procedures, and science processes

- **2.3.1 Using apparatus, equipment, and computers**
- **2.3.2 Conducting routine experimental operations**
- **2.3.3 Gathering data**
- **2.3.4 Organizing and representing data**
- **2.3.5 Interpreting data**

2.4 Investigating the natural world

- **2.4.1 Identifying questions to investigate**
- **2.4.2 Designing investigations**
- **2.4.3 Conducting investigations**
- **2.4.4 Interpreting investigational data**
- **2.4.5 Formulating conclusions from investigational data**

2.5 Communicating

- **2.5.1 Accessing and processing information**
- **2.5.2 Sharing information**

Perspectives

3.1 Attitudes towards science, mathematics, and technology

3.1.1 Positive attitudes toward science, mathematics, and technology

3.1.2 Skeptical attitudes towards use of science and technology

3.2 Careers in science, mathematics and technology

3.2.1 Promoting careers in science, mathematics, and technology

3.2.2 Promoting importance of science, mathematics, and technology in non-technical careers

3.3 Participation in science and mathematics by underrepresented groups

3.4 Science, mathematics, and technology to increase interest

3.5 Safety in science performance

3.6 Scientific habits of mind

Appendix B

DOCUMENTS ANALYZED

Science:

Document Type	Document Title/Publisher/Date
Curriculum Guide	*Chapter 75 Curriculum- for Grades K-12.* Texas Education Agency, 1992
	The Chemistry of Matter - Block H. The University of the State of New York. The State Department of Education, 1988
	Core Course Proficiencies: Science. New Jersey Department of Education, 1990
	Curriculum Frameworks for Grades 6-8 Basic Programs- Science and Health (Florida). Florida Department of Education
	Elementary Science Course of Study and Curriculum Guide. Idaho State Department of Education, 1989
	Elementary Science Curriculum Guide. New Jersey Department of Education, 1985
	Elementary Science Syllabus. The University of the State of New York. The State Department of Education, 1992
	Energy: Sources and Issues - Block I. The University of the State of New York. The State Department of Education, 1991
	Energy and Motion - Block G. The University of the State of New York. The State Department of Education, 1990
	Essential Content. Hawaii Department of Education, 1992
	A Guide to Curriculum Planning in Science. Wisconsin Department of Public Instruction, 1990
	Guidelines for Science Curriculum in Washington Schools. Office of the Superintendent for Public Instruction, 1992
	Learner Outcomes - Science. Oklahoma State Department of Education
	Quality Core Curriculum Guides for K-12 Students in Georgia.
	Science. State of Nevada, Department of Education

Science Comprehensive Curriculum Goals- a Model for Local Curriculum Development. Oregon State Department of Education, 1989

Science Curriculum Guide- Section 3, Science. Idaho State Department of Education, 1989

Science Framework for Grades K-12. California State Board of Education, 1990

State Goals and Sample Learning Objectives- Biological and Physical Sciences for grades K-12. Illinois State Department of Education

Weather and Climate - Block E. The University of the State of New York. The State Department of Education, 1988

Wyoming Standards of Excellence for Science Education. Wyoming Department of Education, 1989

Textbooks

Earth Science. Merrill, 1993

HBJ Science- for grade 4. Harcourt Brace Jovanovich, 1989

Life Science. Merrill, 1993

Modern Physics. Holt, Rinehart and Winston, 1992

Physical Science- for grade 8. Merrill, 1993

Science Horizons. Silver Burdett and Ginn, 1991

Science in Your World. McMillan / McGraw Hill, 1991

University Physics. Addison Wesley, 1992

Mathematics:

Curriculum Guide

A Guide to Curriculum Development in Mathematics. State of Connecticut Board of Education, 1981

A Guide to Curriculum Planning in Mathematics. Wisconsin Department of Public Instruction, 1991

Curricular Frameworks. Florida Department of Education, 1990

Essential Content. Hawaii Department of Education, 1992

Grades K-6 Mathematics Program Guide. Hawaii Department of Education, 1978

Guidelines for K-8 Mathematics Curriculum. Office of the Superintendent for Public Instruction (Washington), 1992

Guidelines for Grades 9-12. Washington State Department of Education, 1985

Kentucky Mathematics Framework. Kentucky Department of Education, 1989

Kentucky Program of Studies- Mathematics. Kentucky Department of Education, 1989

Maryland State Performance Program. Maryland State Department of Education, 1983

Mathematics: Arkansas Public School Course Content Guide. Arkansas State Board of Education

Mathematics Education: An Empowerment. Wyoming Department of Education

Mathematics Framework for California Public Schools. California Department of Education, 1992

Mathematics- A Maryland Curriculum Framework. Maryland State Department of Education, 1983

Mathematics-GRADES 1-8. Arkansas Department of Education

Michigan Essential Goals and Objectives for Mathematics Education. Michigan State Board of Education, 1990

Model Core Curriculum Outcomes and Position Statement on Core Curriculum. Michigan State Board of Education & Michigan Department of Education, 1991

Restructuring the Curriculum. Hawaii Department of Education, 1992

South Carolina Mathematics Framework, Field Review Draft. South Carolina Department of Education, 1992

State Goals for Learning Objectives. Illinois State Board of Education, Department of School Improverment Services, 1985

Student Outcomes for the Foundation Program. State of Hawaii

Wyoming Standards of Excellence for Mathematics Education. Wyoming Department of Education, 1990

Textbooks

Exploring Mathematics (Grade 8). Scott Foresman, 1991

Mathematics - 4. Addison Wesley, 1993

Mathematics-Exploring Your World (Grade 4). Silver Burdett and Ginn, 1992

Mathematics in Action (Grade 4). MacMillan / McGraw Hill, 1992

Algebra: Structure and Method Book 1. Houghton Mifflin Co., 1994

Mathematics (Grade 8). Addison-Wesley, 1993

Mathematics in Action (Grade 8). MacMillan / McGraw Hill, 1992

Mathematics- Exploring Your World (Grade 8). Silver Burdett and Ginn, 1992

Calculus with Analytic Geometry. John Wiley and Sons, 1992

Calculus- A Graphing Approach. Addison-Wesley, 1993

Appendix C

MEANS, PROPORTIONS, AND STANDARD ERRORS FOR TEACHER DATA

Exhibit 29.

Number of periods grade 8 mathematics teachers covered specific topics.

	U.S. Regular		U.S. Pre-Algebra		U.S. Algebra		Germany		Japan	
	Mean	Standard Error	Mean	Standard Error	Mean	Standard Error	Mean	Standard Error	Mean	Standard Error
Whole Numbers	7.5	0.53	8.0	0.72	4.6	0.60	2.4	0.49	1.3	0.27
Fractions & Decimals	24.0	1.43	22.4	1.97	14.8	1.68	16.4	1.83	2.4	0.43
Other Numbers and Number Concepts	10.1	0.56	9.8	0.46	11.4	0.76	5.4	0.86	2.4	0.45
Number Theory and Counting	7.7	0.43	7.4	0.70	6.8	0.49	1.9	0.36	0.8	0.14
Estimation and Number Sense	5.5	0.29	6.1	0.36	4.1	0.39	2.6	0.34	1.5	0.23
Measurement Units	6.2	0.44	6.5	0.68	4.0	0.41	4.4	0.60	1.3	0.22
Perimeter, Area and Volume	7.4	0.58	6.8	0.61	4.9	0.72	13.9	1.36	2.2	0.42
Measurement Estimation and Error	2.2	0.26	2.6	0.35	1.5	0.28	1.6	0.25	2.3	1.24
2-D Geometry	6.0	0.53	5.4	0.45	6.7	0.56	10.4	1.00	18.4	1.11
Geometry Transformations	2.2	0.32	2.1	0.33	1.5	0.37	3.6	0.70	1.6	0.33
Congruence and Similarity	4.1	0.41	2.6	0.38	2.0	0.44	7.0	0.93	31.3	1.80
3-D Geometry, Vectors, Constructions	2.2	0.32	2.7	0.82	1.1	0.26	6.6	0.74	1.7	0.36
Proportionality Concepts	4.0	0.17	3.7	0.19	4.1	0.27	2.5	0.34	3.3	0.52
Proportionality Problems	4.8	0.30	4.2	0.39	4.9	0.43	4.2	0.51	5.2	0.59
Slope and Trigonometry	0.8	0.34	1.2	0.39	3.3	0.50	1.1	0.40	3.0	0.36
Linear Interpolation and Extrapolation	0.3	0.06	0.5	0.21	1.3	0.43	0.4	0.17	4.3	0.82
Patterns, Relations, and Functions	2.4	0.39	2.5	0.49	6.4	0.68	11.1	1.16	16.8	0.99
Equations and Formulas	7.9	0.59	8.9	0.72	16.1	0.86	20.9	1.16	25.2	1.79
Data Representation and Analysis	3.9	0.44	3.6	0.61	2.7	0.38	2.3	0.45	3.0	0.54
Uncertainty and Probability	2.3	0.30	2.4	0.43	2.4	0.53	1.6	0.35	0.0	0.00
Sets and Logic	1.4	0.23	1.1	0.38	3.7	0.42	1.3	0.35	0.4	0.11
Other Content	9.5	0.68	11.7	1.04	13.0	0.81	8.0	0.88	11.2	1.06

Exhibit 30.

Number of periods grade 8 science teachers covered specific topics.

	U.S. Earth		U.S. Life		U.S. Physical		U.S. General		Germany		Japan	
	Mean	Standard Error	Mean	Standard Error	Mean	Standard Error	Mean	Standard Error	Mean	Standard Error	Mean	Standard Error
Earth Features	19.3	1.37	2.7	0.96	2.0	0.67	9.4	1.13	3.2	0.69	3.1	0.30
Earth Processes	10.8	0.58	1.7	0.79	1.0	0.33	5.4	1.10	1.9	0.48	5.7	0.59
Earth in the Universe	9.3	1.10	1.7	0.60	1.4	0.39	5.6	0.75	2.3	0.48	2.3	0.52
Div., Org. & Struc. of Living Things	1.9	0.73	19.0	2.45	1.0	0.43	5.7	1.41	17.6	1.50	15.4	0.75
Life Processes & Systems	1.3	0.52	16.8	2.37	0.7	0.26	5.3	1.26	14.5	1.54	18.0	0.90
Life Spirals, Genetic Continuity & Div.	2.5	0.65	15.6	3.02	1.3	0.58	8.4	1.43	9.7	1.21	2.8	0.41
Interactions of Living Things	1.7	0.42	4.6	3.50	0.3	0.10	3.4	0.54	8.6	1.06	1.6	0.24
Human Biology & Health	1.8	0.58	11.5	3.12	0.5	0.26	4.2	1.10	10.1	1.38	10.6	0.71
Matter	4.7	0.27	3.7	1.07	9.3	0.79	5.6	0.68	3.9	0.63	9.4	0.67
Structure of Matter	6.5	0.43	4.0	0.92	9.7	0.91	6.2	0.71	4.0	0.66	7.8	0.50
Energy & Physical Processes	4.6	0.53	1.9	0.72	13.4	1.67	8.9	0.91	12.5	1.73	9.5	0.78
Physical Transformations	4.3	0.56	1.8	0.53	10.2	0.54	4.1	0.46	3.2	0.43	2.0	0.29
Chemical Transformations	3.8	0.50	3.9	1.43	10.1	0.93	4.7	0.77	2.3	0.47	11.8	0.65
Forces & Motion	2.1	0.38	0.3	0.18	8.6	1.00	3.8	0.72	3.7	0.59	0.2	0.08
Science, Technology & Society	3.9	0.35	3.3	0.99	5.2	0.57	3.5	0.57	1.2	0.34	0.4	0.10
History of Science & Technology	2.4	0.30	2.4	0.64	3.3	0.49	2.4	0.31	3.6	0.50	2.0	0.27
Environmental & Resource Issues	5.9	0.48	1.4	0.54	3.0	0.43	4.0	0.62	6.1	0.65	1.5	0.18
Nature of Science	5.3	0.31	4.7	0.85	5.7	0.50	4.7	0.59	4.1	0.47	1.0	0.19

Exhibit 31.

The five topics emphasized most by grade 8 mathematics teachers.

	U.S. Non-Algebra		U.S. Algebra		Germany		Japan	
	Mean Proportion	Standard Error	Mean Proportion	Standard Error	Mean Proportion	Standard Error	Mean Proportion	Standard Error
Most Emphasized Topic	0.16	0.01	0.13	0.01	0.16	0.01	0.22	0.01
Second Most Emphasized Topic	0.07	0.00	0.11	0.01	0.11	0.01	0.18	0.01
Third Most Emphasized Topic	0.07	0.00	0.10	0.01	0.10	0.01	0.13	0.01
Fourth Most Emphasized Topic	0.06	0.00	0.09	0.01	0.08	0.01	0.12	0.01
Fifth most Emphasized Topic	0.06	0.00	0.05	0.00	0.08	0.01	0.08	0.01

Exhibit 32.

The five topics emphasized most by grade 8 science teachers.

	U.S. Earth		U.S. Life		U.S. Physical		U.S. General		Germany		Japan	
	Mean Proportion	Standard Error	Mean Proportion	Standard Error	Mean Proportion	Standard Error	Mean Proportion	Standard Error	Mean Proportion	Standard Error	Mean Proportion	Standard Error
Most Emphasized Topic	0.21	0.01	0.20	0.02	0.16	0.02	0.10	0.01	0.16	0.02	0.17	0.01
Second Most Emphasized Topic	0.12	0.01	0.17	0.02	0.12	0.01	0.10	0.01	0.13	0.01	0.14	0.01
Third Most Emphasized Topic	0.11	0.01	0.15	0.02	0.11	0.01	0.09	0.02	0.10	0.01	0.12	0.01
Fourth Most Emphasized Topic	0.07	0.01	0.06	0.02	0.11	0.01	0.07	0.01	0.06	0.01	0.09	0.01
Fifth most Emphasized Topic	0.07	0.01	0.04	0.01	0.11	0.01	0.06	0.01	0.05	0.01	0.09	0.01

Exhibit 34.

Pedagogical approaches grade 8 mathematics teachers chose in two pedagogical situations.

	U.S.		Japan	
Item 1	%	Standard Error	%	Standard Error
Constructivist	14.7	2.13	17.3	3.11
Deductive	64.5	3.96	27.5	4.55
Inductive	17.5	3.15	38.4	4.91
Text Dominated	10.0	3.10	22.9	3.48
Item 2	%	Standard Error	%	Standard Error
Constructivist	20.5	2.95	24.0	3.58
Deductive	0.7	0.40	8.4	2.58
Inductive	71.7	3.80	65.0	4.37
Text Dominated	9.6	3.16	11.2	2.69

Exhibit 35.

Grade 8 mathematics and science teacher awareness of curriculum-relevant documents.

	U.S. Math		U.S. Science	
	%	Standard Error	%	Standard Error
NCTM/AAAS	85.5	3.44	25.5	2.87
State Curriculum Guide	63.5	4.65	64.2	4.93
District Curriculum Guide	83.8	3.12	78.56	4.26
NAEP Exam Specs	24.0	3.22	15.68	2.80
State Assessment Specs	41.2	3.53	39.35	4.54

Exhibit 36.

Number of hours per week taught by mathematics and science teachers.

	U.S.		Germany		Japan	
	Mean	Standard Error	Mean	Standard Error	Mean	Standard Error
Mathematics	31.21	0.80	22.46	0.44	19.19	0.26
Science	30.85	0.69	23.63	0.34	18.56	0.29

Exhibit 41.

Number of activities used by grade 8 mathematics teachers in one period.

	U.S.		Germany		Japan	
Number of Activities	%	Standard Error	%	Standard Error	%	Standard Error
1	0.7	0.68	1.8	1.28	3.1	1.41
2	0.8	0.83	3.7	2.17	7.3	2.17
3	6.0	2.85	10.8	3.57	10.7	2.64
4	10.4	2.27	28.6	4.73	27.1	3.35
5	20.9	2.46	30.0	4.75	28.8	3.68
6+	61.2	3.43	25.19	5.01	22.9	3.28

Exhibit 42.
Cumulative distribution of the number of activities used by grade 8 mathematics teachers in one period.

	U.S.		Germany		Japan	
Number of Activities	%	Standard Error	%	Standard Error	%	Standard Error
1	0.7	0.68	1.8	1.28	3.1	1.41
2	0.8	0.83	3.7	2.17	7.3	2.17
3	6.0	2.85	10.8	3.57	10.7	2.64
4	10.4	2.27	28.6	4.73	27.1	3.35
5	20.9	2.46	30.0	4.75	28.8	3.68
6	23.3	2.74	16.8	4.04	12.9	3.14
7	20.8	3.39	6.5	2.78	5.5	1.92
8	10.2	2.16	1.9	1.48	3.8	1.62
9	3.9	1.99	0.0	0.00	0.0	0.00
10	2.5	1.18	0.0	0.00	0.0	0.00
11	0.5	0.53	0.0	0.00	0.8	0.81

Exhibit 43.
Cumulative distribution of the number of activities used by grade 8 science teachers in one period.

	U.S.		Germany		Japan	
Number of Activities	%	Standard Error	%	Standard Error	%	Standard Error
1	1.92	1.47	1.7	1.05	2.4	1.23
2	3.42	1.62	3.5	1.78	5.1	1.67
3	15.47	3.2	18.9	3.98	24.0	3.58
4	26.71	3.42	22.2	4.03	25.2	3.74
5	15.98	2.79	35.9	5.27	27.5	3.98
6	14.88	3.22	12.4	3.34	13.4	3.13
7	12.46	2.97	2.2	1.22	1.6	1.12
8	3.69	2.65	3.2	1.61	0.0	0.00
9	3.84	1.02	0.0	0.00	0.0	0.00
10	0.35	0.28	0.0	0.00	0.0	0.00
11	1.27	0.70	0.0	0.00	0.9	0.04

Exhibit 44.
Time grade 8 mathematics teachers reported having students do exercises or homework during class.

	U.S.		Japan		Germany	
	Minutes	Standard Error	Minutes	Standard Error	Minutes	Standard Error
Paper & Pencil Exercises	7.38	0.46	13.38	0.81	9.07	0.78
Do Homework in Class	7.03	0.87	0.66	0.12	0.19	0.15

Appendix D

List of Exhibits

Endnotes

[1] Schmidt, W.H. et al. 1996 *Characterizing Pedagogical Flow: An Investigation of Mathematics and Science Teaching in Six Countries.* Dordrecht, the Netherlands: Kluwer Academic Press.

[2] Schmidt, W.H.; McKnight, C.E.; Valverde, G.A.; Houang, R.T. and Wiley, D.E. 1996. *Many Visions, Many Aims: A Cross-National Investigation of Curricular Intentions in School Mathematics.* Dordrecht, the Netherlands: Kluwer Academic Press — and: Schmidt, W.H.; Raizen, S.A.; Britton, E.D.; Bianchi, L.J. and Wolfe, R.G. 1996 *Many Visions, Many Aims: A Cross-National Investigation of Curricular Intentions in Science Education.* Dordrecht, the Netherlands: Kluwer Academic Press.

[3] National Council of Teachers of Mathematics. 1989. *Curriculum and Evaluation Standards for School Mathematics.* Reston, VA: National Council of Teachers of Mathematics.

[4] Project 2061: American Association for the Advancement of Science. 1993. *Benchmarks for Science Literacy.* New York: Oxford University Press.

[5] National Research Council. 1995. *National Science Education Standards.* Washington, DC: National Academy Press.

[6] For example, see Blank, R.K., and Pechman, E.M. May, 1995. *State Curriculum Frameworks in Mathematics and Science: How Are They Changing Across the States?* Washington, DC: Council of Chief State School Officers.

[7] Schmidt, W.H.; McKnight, C.E.; Valverde, G.A.; Houang, R.T. and Wiley, D.E. 1996. *Many Visions, Many Aims: A Cross-National Investigation of Curricular Intentions in School Mathematics.* Dordrecht, the Netherlands: Kluwer Academic Press — and: Schmidt, W.H.; Raizen, S.A.; Britton, E.D.; Bianchi, L.J. and Wolfe, R.G. 1996 *Many Visions, Many Aims: A Cross-National Investigation of Curricular Intentions in Science Education.* Dordrecht, the Netherlands: Kluwer Academic Press.

[8] National Council of Teachers of Mathematics. 1989. *Curriculum and Evaluation Standards for School Mathematics.* Reston, VA: National Council of Teachers of Mathematics.

[9] National Council of Teachers of Mathematics. 1989. *Curriculum and Evaluation Standards for School Mathematics.* Reston, VA: National Council of Teachers of Mathematics.

[10] Textbooks do not formally limit teachers' choices, but in the demanding world of real classrooms, teachers must often choose not what they think best but rather what they have time and energy enough to use. In that context, textbooks often become curricula by default. Policy documents may set out the "macro" level of

goals, objectives, and sequences for a grade or several grades. Textbook adoption or availability may shape their implementation. Specifying suggested or required parts to cover in adopted textbooks can further shape implementation, whether this happens at the local or state level. Such official policies are rarely, if ever, detailed enough to guide day-to-day activities during mathematics and science class periods.

[11] Teachers may plan instruction by choosing parts, planning emphases, and supplementing textbooks' materials. The result for hard-pressed U.S. teachers is often essentially an automatic (non-)decision such as those made by other professionals, often with higher status and better pay. It is similar, for example, to choosing to use existing software for word processing, rather than to write a new computer program.

[12] Simon, H.A. 1975. *Models of Man, Social and Rational: Mathematical Essays on Rational Human Behavior in a Social Setting.* New York: Wiley.

[13] Schmidt, W.H. et al. 1996 *Characterizing Pedagogical Flow: An Investigation of Mathematics and Science Teaching in Six Countries.* Dordrecht, the Netherlands: Kluwer Academic Press.

[14] In fact, a team of researchers examining data from the Second IEA Mathematics Study, noted that U.S. teachers held a *variety of logically inconsistent views about mathematics education,* they speculated that this was at least in part due to teachers attempting to take a variety of sources of information and policy seriously — despite their inconsistencies. (see: Sosniak, L.A., Ethington, C.A. and Varelas, M. 1991 Teaching Mathematics without a coherent point of view. *Journal of Curriculum Studies.* 23(2) 119-131.

[15] Many have pointed out this characteristic of educational policy in the U.S. See, for example: Kirst, M.S. 1994 "The Politics of Nationalizing Curricular Content." *American Journal of Education* 102, 383-393; also — Fuhrman, S.H. and Malen, B (eds.) 1991. *The Politics of Curriculum and Testing: The 1990 Yearbook of the Politics of Education Association.* London: The Falmer Press.

[16] Federalism is inherent in the U.S. political structure. We are a union of separate states that grant some of their powers to our federal government to more effectively provide for our common good. That federalism permeates our approach to education. Educationally, we are at the same time one nation, 50 states, and thousands of school districts.

[17] Historically, our federalism has varied in how loosely or tightly united our states have been. It varies among different areas of the common good —mutual defense, health care, safety regulations, providing for the needy or the aged, and, above all, in providing for the common education for all to which our nation is committed.

[18] This educational federalism surely relates to the more general U.S. federalist political structure. However, this federalism of visions is one source of our visions' splintering in our conceptions and policies that guide science and mathematics education. Different roots, other sources, explain why school mathematics and science widely replicate fragmentation into sequences of narrow, specialized goals. Either source would result in partly comparable, partly disparate fragmentary visions; together they certainly do.

[19] These official participants aggregate in a loose federalism providing guiding visions for school mathematics and science. How much diversity exists among these guiding visions depends on the state of public debate and consensus at any point.

[20] This vision, these combined goals, are not matters of official, *de jure* national statement. They are the concomitants of our national educational "yield," the set of all the things we do to produce the outcomes we measure by national samples in assessments such as NAEP and TIMSS.

[21] In 1989, the U.S. Governors Alliance (NGA) and the President met in an "education summit" in Charlottesville, VA. They declared after the summit that "the time has come, for the first time in U.S. history, to establish clear national performance goals, goals that will make U.S. internationally competitive." (White House press release, Office of the Press Secretary. Feb. 26, 1990. page 1). The White House press release

sometime later articulated the "six national education goals" arising from that educational summit, including "Goal 4," that "By the year 2000, U.S. students will be the first in the world in science and mathematics achievement." (page 4).

[22] For example, this was done in such statements as: "...Governors will work within their own sites to develop strategies for restructuring their education systems in order to achieve the goals. Because states differ from one another, each state will approach this in a different manner. The President and the Governors will work to support these state efforts, and to recommend steps that the federal government, business, and community groups should take to help achieve these national goals." (White House press release, Office of the Press Secretary. Feb. 26, 1990. page 6).

[23] An early step in trying to achieve these goals was creating the National Education Goals Panel (NEGP) in July 1990. The panel was charged with "measuring our progress over ten years toward achieving the six National Education Goals." (National Education Goals Panel. [1990]. *The National Education Goals Report: Building A Nation of Learners.* Page iii). The legislation authorizing these actions was entitled "Goals 2000," a reference to the goals of the first NGA education summit. From the beginning national goals were linked to assessment and testing. In 1991, the U.S. Congress and NEGP created the National Council on Education Standards and Testing (see Blank, R.K., and Pechman, E.M. May, 1995. *State Curriculum Frameworks in Mathematics and Science: How Are They Changing Across the States?* Washington, DC: Council of Chief State School Officers). In forming this council, official national concern moved from goals to include testing and setting standards for student achievements. These actions conspicuously and deliberately disavowed any effort at mandatory, official national curricula.

[24] The U.S. National Council of Teachers of Mathematics (NCTM) issued a major set of standards for mathematics curricula in 1989. (National Council of Teachers of Mathematics. 1989. *Curriculum and Evaluation Standards for School Mathematics.* Reston, VA: National Council of Teachers of Mathematics.) The NCTM later issued standards for assessment and teacher preparation. Also in 1989, the American Association for the Advancement of Science published a discussion of scientific literacy (including mathematics) for high school graduates. A more formal report followed in 1993 (Project 2061: American Association for the Advancement of Science. 1993. *Benchmarks for Science Literacy.* New York: Oxford University Press.) The latter contained 'benchmarks' which were not "offering a standard curriculum to be adopted locally" but, instead, provided "educators in every state and school district with a powerful tool to use in fashioning their own curricula" (page vii). Clearly, this report implicitly recognized a federalism of visions and explicitly rejected the idea of a national curriculum. More recently, the National Research Council of the National Academy of Sciences issued yet another, more official set of standards for science education (National Research Council. 1995. *National Science Education Standards.* Washington, DC: National Academy Press). This group includes representatives of the sciences and mathematics from throughout the U.S. and has the official responsibility to advise the U.S. government on issues of science policy.

[25] Blank, R.K., and Pechman, E.M. May, 1995. *State Curriculum Frameworks in Mathematics and Science: How are They Changing Across the States?* Washington, DC: Council of Chief State School Officers.

[26] For a recent discussion see A.C. Lewis, *Phi Delta Kappan,* June 1995. One author described the situation in 1995 by writing, "..America is awash with competing schemes to save the schools. Some of these resemble Goals 2000, many do not, and other seek to save the schools from Goals 2000 and similar reforms" (D. Cohen. *What standards for national standards?* Phi Delta Kappan, June 1995, page 752).

[27] In 1996, the governors and the CEOs of 50 major companies met and again affirmed related goals. The hosts for this 1996 "National Education Summit" of the National Governors' Association (NGA) were IBM's Chair and CEO and the governors of Wisconsin and Nevada (chair and vice-chair, respectively, of the NGA). The announced goal of the Summit was, ".. to build commitment among participants for prompt actions that will help states and communities build consensus, develop and implement high academic standards, assessments

and accountability" (Briefing book for the 1996 NGA Education Summit, distributed on the World Wide Web, March, 1996). Clearly the emphasis was not on the federal government as a key participant in the reform process (although the President spoke at the Summit) but, rather, on bringing some coherence to our educational federalism. The concern for standards and assessment was at the forefront.

[28] Lewis, A.C. *Phi Delta Kappan,* June 1995, page 747.

[29] Cohen, D. "What Standards For National Standards?" *Phi Delta Kappan,* June 1995, page 752, where he wrote, "..[F]rom the state or local perspective [Goals 2000] seems like just another voice in a... babel of reform proposals."

[30] The same author went on to write, "Goals 2000 has been a single locus for ideas about national school reform. But the action in school reform lies elsewhere — in hundreds or thousands of groups that advocate specific reforms, in hundreds more agencies that tend the school system and worry about reform, in the 50 states, in about 15,000 school systems, and in scores of thousands of schools" (D. Cohen. "What standards for national standards?" *Phi Delta Kappan,* June 1995, page 752). This statement is about reform generally, not mathematics and science education reform particularly. However, it clearly applies to them.

[31] Official national actions, subsumed mostly under the general heading of the Goals 2000, have been a lightning rod of controversy. The national role is alternately redefined, de-emphasized, and emphasized anew. Why does the need for a wider consensus on educational visions, goals, intentions, and standards return so often to the forefront of discussion? One editorial recently characterized the situation by stating, "Why do we need the standards movement to succeed? Because right now we have a system that works against itself....a system whose vague and unarticulated goals make it more difficult to extend a helping hand to those youngsters who need it most..." (*American Educator,* Spring brings new life to the standards movement, Spring 1996, page 7). Whether linked to state, local, or federal efforts, improving our aggregate educational system — here, in mathematics and science education — continues to press our public thought. Even with no explicit force or forum driving U.S. to continue, we appear unable to abandon the search for intelligent, shared change.

[32] Allison, Graham T. 1971. *Essence of Decision: Explaining the Cuban Missile Crisis.* Glenview, IL: Scott, Foresman and Company.

[33] Allison, Graham T. op. cit.

[34] Allison, Graham T. op. cit.

[35] Allison, Graham T. op. cit. p. 67

[36] Allison, Graham T. op. cit. p. 82

[37] Simon, H.A. 1957. *Models of Man, Social and Rational: Mathematical Essays on Rational Human Behavior in a Social Setting.* New York: Wiley.

[38] Edelson, Paul J. *Socrates on the Assembly Line: The Ford Foundation's Mass Marketing of Liberal Adult Education.* Paper presented in the Annual Conference of the Midwest History of Education Society. Chicago, IL, October 18-19, 1991 (page 19).

[39] Eli Whitney, though best known for inventing the cotton gin, had perhaps a greater innovation when he pioneered mass production in manufacturing methods. Fearing war with France, the U.S. government in 1798 gave Whitney a contract of $134,000 to produce and deliver 10,000 muskets. In fulfilling this contract, Whitney successfully applied his "Uniformity System" of manufacturing interchangeable components. Whitney convinced skeptics (including President John Adams) of the method's significance by randomly selecting parts and fitting them together into a complete, working musket assembly. Whitney's work on the muskets continued 10 years and during that time he established his system as the seminal breakthrough behind

the modern assembly line. He demonstrated that machine tools manned by ordinary workers (not gunsmith craftsmen) could produce standardized parts to precise specifications, and that any part could be used as a component of any musket. See, for example, C.M. Green, *Eli Whitney and the Birth of American Technology* (1956).

[40] Eisner, Elliot W. "Standards for American Schools: Help or Hindrance?" *Phi Delta Kappan,* June, 1995, page 759.

[41] Gagné and Briggs, L. 1979. *Principles of Instructional Design, 2nd ed.* New York: Holt, Rinehart and Winston.

[42] Callahan, R. 1962. *Education and the Cult of Efficiency.* Chicago, IL: University of Chicago Press.

[43] Bruner, J.S. 1977. *The Process of Education.* Cambridge, MA: Harvard University Press, p. 13.

[44] Schmidt, W.H. et al. 1996 *Characterizing Pedagogical Flow: An Investigation of Mathematics and Science Teaching in Six Countries.* Dordrecht, the Netherlands: Kluwer Academic Press.

[45] See for example: Stevenson, H.W. and Stigler, J.W. 1992. *The Learning Gap.* New York: Summit Books; Prawat, R.S. 1989. "Teaching for Understanding: Three Key Attributes." *Teaching and Teacher Education* 5(4), 315-328; and Steffe, L.P. and Gale, J. 1995. *Constructivism in Education.* Hillsdale, NJ: Lawrence Erlbaum Associates.

[46] Assembly line techniques were first used in 1798 and, while common education was pursued in North America from colonial times, extensive mass education leaped with the expansion of the U.S. in the nineteenth century.